This series aims to report new developments in mathematical research and teaching - quickly, informally and at a high level. The type of material considered for publication includes:

1. Preliminary drafts of original papers and monographs
2. Lectures on a new field, or presenting a new angle on a classical field
3. Seminar work-outs
4. Reports of meetings, provided they are
   a) of exceptional interest or
   b) devoted to a single topic.

Texts which are out of print but still in demand may also be considered if they fall within these categories.

The timeliness of a manuscript is more important than its form, which may be unfinished or tentative. Thus, in some instances, proofs may be merely outlined and results presented which have been or will later be published elsewhere.

Manuscripts should comprise not less than 100 pages.

Publication of *Lecture Notes* is intended as a service to the international mathematical community, in that a commercial publisher, Springer-Verlag, can offer a wider distribution to documents which would otherwise have a restricted readership. Once published and copyrighted, they can be documented in the scientific literature. -

Manuscripts

Manuscripts are reproduced by a photographic process; they must therefore be typed with extreme care. Symbols not on the typewriter should be inserted by hand in indelible black ink. Corrections to the typescript should be made by sticking the amended text over the old one, or by obliterating errors with white correcting fluid. Authors receive 75 free copies.

The typescript is reduced slightly in size during reproduction; best results will not be obtained unless the text on any one page is kept within the overall limit of 18 x 26.5 cm (7 x 10½ inches). The publishers will be pleased to supply on request special stationery with the typing area outlined.

Manuscripts in English, German or French should be sent to Prof. Dr. A. Dold, Mathematisches Institut der Universität Heidelberg, 69 Heidelberg/Germany, Tiergartenstraße or Prof. Dr. B. Eckmann, Eidgenössische Technische Hochschule, CH-8006 Zürich/Switzerland.

Lecture Notes in Physics

Bisher erschienen/Already published

Vol. 1: J. C. Erdmann, Wärmeleitung in Kristallen, theoretische Grundlagen und fortgeschrittene experimentelle Methoden. II, 283 Seiten. 1969. DM 20,–

Vol. 2: K. Hepp, Théorie de la renormalisation. III, 215 pages. 1969. DM 18,–

Vol. 3: A. Martin, Scattering Theory: Unitarity, Analytic and Crossing. IV, 125 pages. 1969. DM 14,–

Vol. 4: G. Ludwig, Deutung des Begriffs physikalische Theorie und axiomatische Grundlegung der Hilbertraumstruktur der Quantenmechanik durch Hauptsätze des Messens. XI, 469 Seiten. 1970. DM 28,–

Vol. 5: M. Schaaf, The Reduction of the Product of Two Irreducible Unitary Representations of the Proper Orthochronous Quantummechanical Poincaré Group. IV, 120 pages. 1970. DM 14,–

Vol. 6: Group Representations in Mathematics and Physics. Edited by V. Bargmann. V, 340 pages. 1970. DM 24,–

Vol. 7: R. Balescu, J. L. Lebowitz, I. Prigogine, P. Résibois, Z. W. Salsburg, Lectures in Statistical Physics. V, 181 pages. 1971. DM 18,–

Vol. 8: Proceedings of the Second International Conference on Numerical Methods in Fluid Dynamics. Edited by M. Holt. IX, 462 pages. 1971. DM 28,–

Vol. 9: D. W. Robinson, The Thermodynamic Pressure in Quantum Statistical Mechanics. V, 115 pages. 1971. DM 14,–

Vol. 10: J. M. Stewart, Non-Equilibrium Relativistic Kinetic Theory. III, 113 pages. 1971. DM 14,–

Vol. 11: O. Steinmann, Perturbation Expansions in Axiomatic Field Theory. III, 126 pages. 1971. DM 14,–

Lecture Notes in Mathematics

A collection of informal reports and seminars
Edited by A. Dold, Heidelberg and B. Eckmann, Zürich

274

I. Bucur, J. Giraud, N. Goodman and
J. Myhill, L. Illusie, J. Lambek,
D. S. Scott, M. Tierney

Toposes, Algebraic Geometry and Logic

Dalhousie University, Halifax, January 16–19, 1971

Edited by F. W. Lawvere, Matematisk Institut, Aarhus/Danmark

Springer-Verlag
Berlin · Heidelberg · New York 1972

AMS Subject Classifications (1970): 02 C 15, 02 C 20, 02 K 05, 02 K 10, 06 A 23, 14 A 20, 14 A 99, 14 L 99, 18 A 15, 18 B 05, 18 D 15, 18 F 10, 18 F 20

ISBN 978-3-540-05920-2 Springer-Verlag Berlin · Heidelberg · New York

Offsetdruck: Julius Beltz, Hemsbach/Bergstr.

Preface

This volume partially reports a conference on "Connections between Category Theory and Algebraic Geometry & Intuitionistic Logic" held at Halifax, Nova Scotia, Canada, January 16-19, 1971 under the sponsorship of Dalhousie University.*)
Of the seventy mathematicians participating, eight delivered addresses which are represented by the seven articles and the introduction included here. Many more participants contributed essentially to the conference by giving more or less informal lectures and by taking part in the lively mathematical discussions. Because the University administration had just refused to renew my contract due to my political activities, a ninth invited speaker declined to deliver his address, thus joining in the protest which was supported by a majority of the participants. The tension created by the administration's action failed however to dampen the scientific enthusiasm within the conference. Rather the conference concentrated and sharpened the development whereby two previously unrelated trends in modern mathematics are now each applying concepts and methods developed by the other.

Arnold J. Tingley and his staff at Dalhousie University should be thanked here for their tireless work in preparing the conference and this volume.

F.W.Lawvere

---

*) Research partly supported by National Research Council.

TABLE OF CONTENTS

LIST OF PARTICIPANTS

Banachewski, B.
Bass, H.
Beck, J.
Benabou, J.
Bishop, E.R.
Bitterlich, W.
Booth, P.
Boyter, M.
Bucur, I.
Bunge, M.
Butler B.
Brinkmann, H.B.
Caneau, F.
Cooper, E.
Day, A.
Day, B.J.
Diaconescu, R.
Dubuc, E.
Duskin, J.
Dyck, Z.
Fakir, S.
Fekete, A.E.
Felscher, W.
Finke, G.
Freyd, P.
Gerson, M.
Gildenhuys, D.
Giraud, J.
Goodman, N.
Heath, P.
Howlett, Ch.
Illusie, L.
Joyal, A.
Kelly, M.

Lambek, J.
Lawvere, F.W.
Leisenring, A.
Linton, F.
Manes, E.
Matheson, J.S.
Mendelson, E.
Mitchell, B.
Mitchell, W.T.
Nelson, E.
Neumann, G.
Pare, R.
Regoczel, S.
Rishel, T.
Rosenfield, J.E.
Rowe, K.
Schelter, B.
Schlomiuk, D.
Schlomiuk, N.
Schumaker, D.
Scott, D.
Servedio, F.
Söler, F.
Swaminathan, S.
Szabo, F.
Takeuti, G.
Thiébaud, M.
Thompson, A.C.
Tierney, M.
VanOsdol, A.
Venne, M.
Verdier, J.-L.
Volger, H.
Walker, S.
Wakfer, P.
Yasugi, M.

INTRODUCTION

by

F. William Lawvere

The program of investigating the connections between algebraic geometry and "intuitionistic" logic under the guidance of the form of objective dialectics known as category theory was discussed and moved forward at a conference in January 1971 at Halifax, Nova Scotia where seventy mathematicians representing several fields took part. Some of the lectures delivered are reflected in the seven articles in this volume and in the present introduction.

Our own hopes in the success of the above general program were strengthened by initial progress in carrying out a more special program which will be outlined in this introduction. This is the development on the basis of elementary (first-order) axioms of a theory of "toposes" just good enough to be applicable not only to sheaf theory, algebraic spaces, global spectrum, etc. as originally envisaged by Grothendieck, Giraud, Verdier, and Hakim but also to Kripke semantics, abstract proof theory, and the Cohen-Scott-Solovay method for obtaining independence results in set theory. At Rome and Oberwolfach meetings in Spring 1969 I had discussed this program and proposed a set of axioms (essentially theorems 1 and 2 below) which were then shown during my 1969 - 70 collaboration with Myles Tierney to be adequate for all the usual exactness properties of toposes as well as for the construction of sheaf categories and the proof that they are again toposes. That part of our joint work dealing with the continuum hypothesis is detailed in Tierney's article in this volume. We also simplified the original axioms, a process which has been carried further more recently by Chris Juul-Mikkelsen. The proof of the exactness of the associated-sheaf functor has recently been simplified by Peter Freyd, who has also made interesting contributions to the relationship between right-exactness and number theory in a topos. Dana Scott has pointed out that Dedekind-cut sense of "analysis in a topos" reduces to his model for intuitionistic analysis in the "classical" case of sheaves over a non-trivial topological space.

We now understand by a topos any category $\underline{E}$ which is cartesian closed and has a subobject-representor. Thus a topos has a terminal object 1 and cartesian product and exponential functors determined by the adjointness relations

$$\frac{X \longrightarrow Y_1 \times Y_2}{X \longrightarrow Y_1, X \longrightarrow Y_2} \qquad\qquad \frac{X \longrightarrow Y^A}{A \times X \longrightarrow Y}$$

as well as a "truth-value" object Ω satisfying the adjointness relation

$$\frac{X \longrightarrow \Omega}{? \rightarrowtail X}$$

where $? \rightarrowtail X$ refers to an arbitrary equivalence class of monomorphisms into X (i.e. an arbitrary subobject of X). More exactly of course the natural bijections indicated by horizontal lines above are mediated by unities

$$X \xrightarrow{\delta} X \times X \qquad\qquad X \xrightarrow{\lambda} (A \times X)^A$$

$$Y_1 \times Y_2 \xrightarrow{\pi i} Y_i \qquad\qquad A \times Y^A \xrightarrow{\varepsilon} Y$$

in the case of the cartesian closed structure, and by

$$1 \xrightarrow{\text{true}} \Omega$$

in the case of the subobject representor. To explain more precisely the working of the latter, regard any morphism $A \xrightarrow{x} X$ as an element of X "defined over A" (this has its usual sense in the case of algebraic geometry) and for any monomorphism $S \overset{m}{\rightarrowtail} X$ say that

$$x \in m \quad \text{iff there exists } \bar{x} \text{ such that} \quad \begin{array}{ccc} A & \dashrightarrow & S \\ & {\scriptstyle x}\searrow & \swarrow{\scriptstyle m} \\ & X & \end{array}$$

$$x = \bar{x}\, m$$

Further, write $true_A$ for the composite (constant) morphism $A \longrightarrow 1 \xrightarrow{true} \Omega$. Then the determining property of Ω is as follows: Given any "propositional function" $X \xrightarrow{\phi} \Omega$ there is a monomorphism $\{X|\phi\}$ with codomain X such that for any $A \xrightarrow{x} X$

$$x \in \{X|\phi\} \text{ iff } x\phi = true_A$$

and conversely <u>every</u> monomorphism with codomain X has a <u>unique</u> "characteristic function" ϕ . (Anders Kock has shown that in fact it suffices to assume the existence of Y^A for the case $Y = \Omega$.)

Briefly we may say that the notion of topos summarizes in objective categorical form the essence of "higher-order logic" (we will explain below how the logical operators become morphisms in a topos) with no axiom of extensionality. This amounts to a natural and useful generalization of set theory to the consideration of "sets which internally develop". In a basic example of algebraic geometry, the development may be viewed as taking place along a parameter which varies over "rings of definition"; in a basic example from intuitionistic logic, the parameter is interpreted as varying over "stages of knowledge". To illustrate we further describe an example and four classes of examples.

The most "abstract" topos is the familiar category $\underline{S}$ of abstract sets and mappings in which, so to speak, the development has been frozen so that morphisms $X \longrightarrow Y$ are entirely determined by what they do to "global" or "external" elements of X , i.e. elements $1 \longrightarrow X$ defined over the terminal object 1. Here of course Y^A is an abstract set which precisely indexes the morphisms $A \longrightarrow Y$ and Ω is a two-element abstract set. There being no development going on in the objects of $\underline{S}$, there is nothing to obstruct the existence of choice functions, and indeed the axiom of choice in a certain sense characterizes models of set theory among toposes. More exactly, Radu Diaconescu has shown that any topos in which epimorphisms split is also generated by the subobjects of 1 and has $\Omega \cong 1 + 1$ (coproduct) and is hence (in view of the results discussed below) a "Boolean-valued model for the elementary theory of the category of sets" if it satisfies an axiom of infinity.

The first class of toposes to be studied as categories was the class of $\underline{E}$ of the form $\underline{E}$ = all $\underline{S}$-valued sheaves on some topological space. In such an example our axioms are verified in terms of the section functor Γ as follows

$$\Gamma(U, Y^X) = \mathrm{Hom}\ (X|U, Y|U)$$

for all sheaves X and Y and all open sets U, and

$$\Gamma(U, \Omega) = \text{Set of all open subsets of } U$$

A related class of toposes are those of the form $\underline{S}^{\underline{P}}$ where $\underline{P}$ is a poset. Here an object X may be analyzed as a family of abstract sets indexed by the elements of $\underline{P}$ and equipped with transition mappings $X_p \longrightarrow X_q$ for $p \leq q$, satisfying the conditions that the transition mapping $X_p \longrightarrow X_p$ is the identity and that the diagram

$$\begin{array}{ccc} X_p & \longrightarrow & X_q \\ & \searrow \quad \swarrow & \\ & X_r & \end{array}$$

of transition mappings commutes whenever $p \leq q \leq r$. A morphism $X \xrightarrow{f} Y$ is any family $X_p \xrightarrow{f_p} Y_p$ of mappings which commutes with the transition mappings

$$\begin{array}{ccc} X_p & \xrightarrow{f_p} & Y_p \\ \downarrow & & \downarrow \\ X_q & \xrightarrow[f_q]{} & Y_q \end{array} \qquad \text{Whenever } p \leq q .$$

Such a category $\underline{S}^{\underline{P}}$ is a topos, with

$(Y^X)_p \stackrel{\sim}{=}$ set of all families f_q as above, except defined only for those q with $p \leq q$,

and

$\Omega_p \stackrel{\sim}{=}$ set of all those subsets S of $\underline{P}$ which satisfy $q \in S \Rightarrow p \leq q$ and

$$q \in S \text{ and } q \leq r \Rightarrow r \in S$$

with the transition mappings $(Y^X)_p \longrightarrow (Y^X)_q$ and $\Omega_p \longrightarrow \Omega_q$ given by restricting. By considering

$$\bar{X}_p = \sum_{s \leq p} X_s$$

we see that any object X is the quotient of an $\bar{X}$ which "increases" (in the sense that the transitions are monomorphisms) modulo an equivalence relation $E_p \subset \bar{X}_p \times \bar{X}_p$ which also increases; this shows the relationship between toposes of the form $\underline{S}^{\underline{P}}$ and the usual model theory for "intuitionistic" logic - namely we need only take account of "equality" in the latter to reduce it to the former. Note that toposes of the form $\underline{S}^{\underline{P}}$ share with toposes of sheaves on topological spaces the property of having non-Boolean internal logic except in the most trivial cases.

Even before sheaf theory or intuitionistic logic mathematicians considered permutation representations of groups, and these give rise also to a distinctive class of toposes. Slightly more generally, let $\underline{G}$ be any Brandt groupoid and let $\underline{S}^{\underline{G}}$ be the category of representations of $\underline{G}$ in abstract sets, with equivariant maps as morphisms. Then

$$(Y^X)_p = Y_p^{X_p} \text{ (\underline{all} mappings)}$$

for each identity p of $\underline{G}$, with the obvious action $f^g = g^{-1}fg$,

and

$$\Omega_p = \text{two element set}$$

with trivial action. In a sense which can be made precise, these are the only toposes which can be defined over $\underline{S}$ in a way which preserves so strictly the topos structure.

By the way of contrast let $\underline{M}$ be any monoid which is not a group (and, for uniqueness of presentation, assume it has no non-identity idempotents). Then in the topos $\underline{S}^{\underline{M}}$ of $\underline{M}$-sets, $\underline{M}$ acting on itself is a canonical generator and we have that

Y^X = set of all equivariant maps $\underline{M} \times X \longrightarrow Y$

Ω = set of all left ideals of $\underline{M}$

both with a natural action of $\underline{M}$.

The three classes just described are of course subsumed under the more general class of toposes having the form of a functor category $\underline{S}^{\underline{C}^{op}}$ where $\underline{C}$ is any small category. Now one of the important features of the theory of toposes is that a great many constructions can be relativised through replacing $\underline{S}$ by an arbitrary base topos $\underline{E}$, and the functor category construction is one of these. For one thing, significance of the condition that a category $\underline{C}$ is "small" is that its "set" of objects and "set" of morphisms have the nature of objects in the base topos and that its domain, codomain and composition operations have the nature of morphisms in the base topos. For another thing, we have from the topology of fiber bundles the idea that a really internal "family" of objects (for example the family of values of a functor) indexed by an object C_o is simply a morphism $X \longrightarrow C_o$, and this idea is if anything even more sensible in a topos. These two observations can be used to define the notion of category $\underline{C}$ in $\underline{E}$ and to define the category of internal $\underline{E}$-valued functors on $\underline{C}$ for any category $\underline{E}$ with finite limits. A topos E has finite limits, which we can prove by constructing either equalizers or intersections of subobjects: if we denote by θ_Y the characteristic function of the diagonal monomorphism $Y \longrightarrow Y \times Y$, then for any pair f_1, f_2 of morphisms $X \longrightarrow Y$, the composite

$$X \xrightarrow{\langle f_1, f_2\rangle} Y \times Y \xrightarrow{\ \ominus_Y\ } \Omega$$

is the characteristic function of the equalizer $\{X | f_1 \ominus_Y f_2\}$ of f_1 with f_2 ; or if we denote by

$$\Omega \times \Omega \xrightarrow{\ \wedge\ } \Omega$$

the characteristic function of $1 \xrightarrow{\langle \text{true}, \text{true}\rangle} \Omega \times \Omega$, we can, given two subobjects of X with characteristic functions ϕ_1 and ϕ_2 , obtain their intersection as $\{X | \phi_1 \wedge \phi_2\}$.

The fact that the internally-defined functor categories (including the special case $\underline{E}/_{\underline{X}}$) are again toposes, as well as the usual exactness properties of toposes such as pullbacks are exact, pushouts of monos are monos, etc. follow from

Theorem 1. For any morphism $X \xrightarrow{f} Y$ in a topos $\underline{E}$, the functor

$$\underline{E}/_X \xleftarrow[f^*]{} \underline{E}/_Y$$

obtained by pulling back along f has a right adjoint Π_f (as well as the obvious left adjoint Σ_f which is just composition with f). As special cases we have, in addition to the exponentiation in $\underline{E}/_Y$, the partial-morphism representor

$$\tilde{B} = \Sigma_{\Omega \to 1} \Pi_{\text{true}} B$$

for any object B of $\underline{E}$ (which satisfies

$$\frac{A \longrightarrow \tilde{B}}{A \longleftarrow\!\!\prec ? \longrightarrow B}$$

or in other words is the right adjoint to the inclusion of $\underline{E}$ into the category with the same objects but with arbitrary partially-defined morphisms) as well as the operations

$$\Omega^X \xrightarrow{\ \forall_X\ } \Omega \qquad \Omega \times \Omega \xrightarrow{\ \Rightarrow\ } \Omega$$

of universal quantification over X and implication which may alternatively be defined as the characteristic maps of the name of true_X and of the equalizer of conjunction with the first projection, respectively. Other forms of universal quantification are, for any $X \xrightarrow{f} Y$, a right adjoint

$$\Omega^X \xrightarrow{\forall_f} \Omega^Y$$

with respect to the natural order of Ω for the operation of composing with f, as well as for any X, an operation

$$\Omega^{\Omega^X} \xrightarrow{\bigcap X} \Omega^X$$

of infinite intersection which forms the "multiplication" part of a triple (dual standard construction, monad) whose functor part is $f \rightsquigarrow \forall_f$ and whose unit part is the singleton map

$$X \xrightarrow[\{\ \}X]{} \Omega^X$$

(which is just the exponential adjoint of δ_X)

It was with use of universal quantification that Chris Juul-Mikkelsen proved the following

Theorem 2. In a topos there exist a strict initial object 0, union of any two subobjects of any object, disjoint sum of any two objects, image factorization of any morphism into epi and mono, equivalence relation generated by any pair $X \rightrightarrows Y$ of morphisms, and coequalizer of any such pair of morphisms.

It follows from theorems 1 and 2 that all epis are coequalizers and that equivalence relations are universal-effective. Moreover the image factorization gives use to various forms

$$\Omega^X \xrightarrow{\exists X} \Omega \quad , \quad \Omega^X \xrightarrow{\exists f} \Omega^Y \quad , \quad \Omega^{\Omega^X} \xrightarrow{\bigcup X} \Omega^X$$

of existential quantification which satisfy appropriate formal relations (rules of inference) but typically "mean actual existence only locally".

There are at least two forms of the idea of a property holding "locally" in a topos. One is intrinsic, and reflects the idea that any epimorphism $S \longrightarrow 1$ is a covering of $\underline{E}$: thus for example a diagram in $\underline{E}$ is said to locally satisfy some property expressed in the language of toposes if there exists S with $S \longrightarrow 1$ epic such that when the diagram is pulled back to $\underline{E}/S$ it has the property in the sense of the topos $\underline{E}/S$. The other notion is with respect to a given $\Omega \xrightarrow{j} \Omega$ which may be thought of as a modal operator to be read "it is j-locally the case that .." and which satisfies the axioms below which in particular mean that j is equivalent to a Grothendieck topology on $\underline{C}$ in the case of a topos of the form $\underline{S}^{\underline{C}^{op}}$. At the Rome and Overwolfach meetings I had pointed out that the usual notion of a Grothendieck topology is equivalent to a single such morphism j ; Tierney showed that the appropriate axioms on j are simply that $jj = j$ and j preserves finite conjunctions.* A subobject $X' \rightarrowtail X$ with characteristic function $X \xrightarrow{\phi} \Omega$ is said to be j-dense if ϕ is j-locally true i.e. if $\phi\ j = \mathrm{true}_X$. The relationships between the two notions of localness arise from the fact that in a subcategory $\underline{E}_j$ of $\underline{E}$ called the category of j-sheaves, a morphism is an epimorphism in the sense of $\underline{E}_j$ iff its image in the sense of $\underline{E}$ is j-dense. By definition, an object Y of $\underline{E}$ is a j-sheaf iff for every j-dense monomorphism $X' \rightarrowtail X$, every morphism $X' \longrightarrow Y$ can be uniquely extended to a morphism $X \longrightarrow Y$. If Y is a j-sheaf and X is any object of $\underline{E}$, Y^X is again a j-sheaf ; thus the full subcategory $\underline{E}_j$ is cartesian closed. Moreover the image Ω_j of j is a sheaf which is a subobject representor for $\underline{E}_j$; thus the category of j-sheaves is again a topos. In sheaf theory an important construction is the associated-sheaf- functor, a left adjoint to the inclusion functor $\underline{E}_j \rightarrowtail \underline{E}$ which is usually

* including the empty conjunction - true.

constructed by a two-step infinite direct-limit procedure. In the absence of (external) infinite direct limits in the axiomatic setting, I found quite another two-step procedure to construct this adjoint: Given X , consider first the image of the canonical map $X \longrightarrow \Omega^X \longrightarrow \Omega_j{}^X$, then form the j-closure of the resulting subobject of the sheaf $\Omega_j{}^X$ —— this closure is the associated sheaf of X . It is easy to see that this associated-sheaf functor preserves products; the important fact that it preserves all finite inverse limite (i.e. that it is left exact) was proved by Tierney using a calculus-of-fractions argument. More recently Freyd has proved the exactness using the facts that every topos has enough injectives, that every injective of $\underline{E}_j$ is injective in $\underline{E}$, and the following

<u>Lemma</u> A diagram

$$\begin{array}{ccc} A & \rightarrowtail & A_1 \\ \downarrow & & \downarrow \\ A_2 & \rightarrowtail & X \end{array}$$

of monomorphisms in a topos is an intersection iff for every injective E , a pair of maps $A_i \longrightarrow E$ has a common extension to X iff it has a common restriction to A .

The inclusions $\underline{E}_j \rightarrowtail \underline{E}$ constitute precisely the full and faithful case of <u>geometrical morphism</u> $\underline{F} \longrightarrow \underline{E}$ between toposes, which means any functor having an exact left adjoint. Another ("surjective") case of geometrical morphism is one for which the adjoint reflects isomorphisms - these are determined by a left exact cotriple (standard construction) on the domain $\underline{F}$. Moreover every geometrical morphism can be uniquely factored into two, the first of which is "surjective" in the sense just described and the second of which is full and faithful. This "image topos" construction applied to a "sections" functor (with "stalks" adjoint)

$$\underline{S}/X \longrightarrow \underline{S}^{\underline{B}^{op}}$$

arising from a topology basis $\underline{B}$ on a set X, gives the usual sheaf category. The latter construction can be relativised, replacing the category of sets by an arbitrary topos.

Carrying out logic, algebra, and analysis within a topos usually requires the axiom of infinity, i.e. the existence of the free unary algebra $N \circlearrowleft s$ on 1 generator. Chris Juul-Mikkelsen has shown in detail that this is equivalent to the existence of free monoid objects, and Peter Freyd has shown that $N \circlearrowleft s$ is characterized by being a fixed point $1 + N \xrightarrow{\approx} N$ and by being the least such in the sense that

$$N \underset{id}{\overset{s}{\rightrightarrows}} N \longrightarrow 1 \quad \text{is a coequalizer}$$

Over a base topos $\underline{S}$ with N there are two important toposes which should be investigated in more detail. One is the category $\underline{A}$ of sheaves on the product space N^N, which has a nice topology basis; besides its importance for intuitionistic analysis, $\underline{A}$ has the property that "analytic spaces" are determined by left exact cotriples in it. The other is sort of proof-theoretic version of the Dedekind-cut construction which yields a topos $R(\underline{S})$ whose truth-values are (in the $\underline{S}$-sense) nonnegative real numbers (including ∞, and with the "reverse" ordering which is convenient for setting up metric spaces as "strong categories"); if $\underline{Q}$ denotes the poset of non-negative rationals, $R(\underline{S})$ is the subcategory of $\underline{S}^Q$ consisting of those X for which $X_q = \varprojlim_{r>q} X_r$. Using Brian Day's theory of convolution, one can extend the usual addition and truncated subtraction of reals to get a (non-cartesian) closed structure on the whole topos $R(\underline{S})$. It should be useful to work out some aspects of differential topology, infinite-dimensional group representations etc. in this setting i.e. "analysis in a topos" making use of the interplay between the external view, to the effect that a topos is a generalized space, and the internal view, to the effect that a topos is a relativized set theory.

References

Freyd	ASPECTS OF TOPOI	(mimeographed)
Kock and Wraith	ELEMENTARY TOPOSES	Aarhus Lecture Notes No. 30, 1971
Lawvere	QUANTIFIERS AND SHEAVES	Actes du Congrès Int. des Math. 1970, tome 1 p. 329
Tierney	SHEAVES AND THE CONTINUUM HYPOTHESIS	this volume

SHEAF THEORY AND THE CONTINUUM HYPOTHESIS

by

Myles Tierney

In this paper I would like to give an account of some joint work of myself and F. W. Lawvere which is concerned with establishing the independence of the Continuum Hypothesis from other axioms of the category of sets. More precisely, we will show that the Continuum Hypothesis does not follow from the axioms of the Elementary Theory of the Category of Sets [3]. Some discussion of the relationship of this result to that of Cohen can be found at the end of the paper.

The exposition will involve several results about topos - i.e. categories of sheaves - and these will be simply stated and used without proof. The reason for this is that, properly speaking, the material of this paper is an application of our axiomatic theory of sheaves - described in [4] and [7]-and should really be so presented. Thus, although we define the concept of topos below, we make no attempt to develop the theory here, since to do so only to prove this result would be to put the cart before the horse.

Since the axioms we use here for the category of sets are somewhat different from those of [3], though equivalent as a group, we should begin by discussing these in some detail. The first, unnumbered, group states that the universe of discourse is a category

---

[1] The author would like to express his gratitude to Dalhousie University for providing the Killam Senior Fellowship for 1969-70, during whose tenure much of the research for this paper was carried out.

$\underline{S}$. The next group, which defines a topos in the sense of [4] is concerned with those properties of sets that can be described by adjointness. Namely, we assume:

Axiom 1. All finite limits and colimits exist.

As is well-known, to satisfy Axiom 1 it is enough to have an initial object 0 , a terminal object 1, a pullback for every pair of maps with common codomain, and a pushout for every pair of maps with common domain.

Axiom 2. $\underline{S}$ is cartesian closed.

That is, for all X and Y there is an object of maps X^Y with the universal property of λ-conversion. Namely,

$$\frac{Z \longrightarrow X^Y}{Z \times Y \longrightarrow X}$$

with the obvious assumptions of naturality. More compactly, though somewhat improperly, we can express this by saying that for all Y , the functor $(\ ) \times Y$ has a right adjoint $(\ )^Y$.

Axiom 3. Subobjects in $\underline{S}$ are representable.

Precisely, there is an object Ω together with a map $1 \longrightarrow \Omega$, which is called "true," such that for any monomorphism $X' \rightarrowtail X$ (such arrows will always denote monomorphisms) there is a unique characteristic map $\phi : X \longrightarrow \Omega$ such that

$$\begin{array}{ccc} X' & \longrightarrow & 1 \\ \big\downarrow & & \big\downarrow \text{true} \\ X & \xrightarrow{\ \phi\ } & \Omega \end{array}$$

is a pullback.

A category $\underline{E}$ satisfying Axioms 1-3 will be called a topos. If $\underline{S}$ is a category of sets, then certainly 1-3 should hold, where X^Y is the set of all maps from Y to X, and Ω is the 2-point set. More generally, suppose $\underline{C}$ is a category in $\underline{S}$, and consider $\underline{S}^{\underline{C}}$ the category of covariant $\underline{S}$-valued functors on $\underline{C}$. Certainly Axiom 1 holds, since limits and colimits in $\underline{S}^{\underline{C}}$ are computed pointwise. If C is an object of $\underline{C}$, let us denote by the same letter the representable functor $(C,-)$ in $\underline{S}^{\underline{C}}$. Then the Yoneda lemma says that for any $F \in \underline{S}^{\underline{C}}$, the natural transformations from C to F are in natural 1-1 correspondence with the set $F(C)$. We use this to determine exponentation and Ω in $\underline{S}^{\underline{C}}$. Namely, if F and G are functors, and if F^G is to exist at all, then we must have

$$\frac{C \longrightarrow F^G}{C \times G \longrightarrow F} .$$

Using this as a definition, one checks that it works. Similarly, if Ω is to exist, then in particular,

$$\frac{C \longrightarrow \Omega}{R \rightarrowtail C}$$

i.e., the value of Ω at C must be the collection of subobjects (in $\underline{S}^{\underline{C}}$) of the representable functor C — these are called cribles — and again one sees that this works. Thus $\underline{S}^{\underline{C}}$ is a topos. In some sense, the characteristic example of a topos is obtained by taking T to be a topological space in $\underline{S}$, and forming Sheaves (T) — the category of $\underline{S}$-valued sheaves on T. Finite limits and colimits are again clear, though one must be a little more careful with the colimits. Since the representable functors in the category of presheaves — i.e., the open sets of T — are themselves sheaves, the same reasoning as in the previous example shows the existence of exponentiation and Ω. Much of the terminology used in the theory of topos comes from this special case. For example, if $U \rightarrowtail 1$ we call U <u>open</u>, a map $1 \longrightarrow X$ is called a <u>global section</u> of X, etc.

In general, if $\underline{E}$ and $\underline{E}_0$ are topos, a <u>geometric morphism</u> of topos $f:\underline{E} \longrightarrow \underline{E}_0$ will be an adjoint pair

$$\underline{E} \underset{f_*}{\overset{f^*}{\rightleftarrows}} \underline{E}_0 \quad ,$$

in which f^* is left adjoint to f_* and preserves finite limits If there exists a geometric morphism from $\underline{E}$ to $\underline{E}_0$, we say $\underline{E}$ is <u>defined over</u> $\underline{E}_0$.

The next axiom, which asserts the existence of a natural number object, is somewhat different from the previous three in that, though it deals with a property of sets expressible in terms of the existence of a certain adjoint, if $\underline{E}$ is defined over $\underline{E}_0$ and $\underline{E}_0$ satisfies the axiom, so does $\underline{E}$.

<u>Axiom 4</u>. (Axiom of infinity). There is an object ω, together with maps $1 \xrightarrow{0} \omega \xrightarrow{s} \omega$, such that for any object X provided with maps $1 \xrightarrow{x_0} X \xrightarrow{t} X$ there exists a unique map $\sigma:\omega \longrightarrow X$ such that

$$\begin{array}{ccccc} & \overset{0}{\nearrow} & \omega & \xrightarrow{s} & \omega \\ 1 & & \downarrow{\scriptstyle\sigma} & & \downarrow{\scriptstyle\sigma} \\ & \underset{x_0}{\searrow} & X & \xrightarrow{t} & X \end{array}$$

commutes.

Though we will not pursue the point here (see [4]), Axioms 1-3 are sufficient for the complete, axiomatic, development of sheaf theory. That is, we can deduce from them the typical exactness properties of set-valued sheaf categories, define topologies, pass to sheaves, etc. One exactness property that we will need later, and might as well state now, is the following: In any category $\underline{E}$

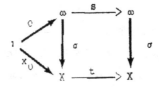

with pullbacks, a map $f:X \longrightarrow Y$ induces a functor

$$\underline{E}/Y \xrightarrow{f^*} \underline{E}/X$$

given by pulling an object over Y back along f to an object over X. It is always true that f^* has a left adjoint Σ_f — just compose with f — but in a topos f^* also has a right adjoint Π_f. This is a much stronger property — it implies, for example, that f^* preserves colimits.

As an example of how one operates with Ω in a topos, let us see what kind of algebraic structure Ω must carry. We already have the map $1 \xrightarrow{\text{true}} \Omega$, and after establishing the fact that $0 \rightarrowtail 1$ is monic, we obtain — as its characteristic map — the map $1 \xrightarrow{\text{false}} \Omega$. Next, there is an operation

$$\wedge : \Omega \times \Omega \longrightarrow \Omega$$

called <u>conjunction</u>, which, by the universal property of Ω, is completely defined by specifying where it takes the value "true." As everyone knows, this is precisely at the pair $\langle\text{true},\text{true}\rangle$, hence $\wedge$ is the characteristic map of the subobject determined by

$$1 \xrightarrow{\langle\text{true},\text{true}\rangle} \Omega \times \Omega \quad .$$

Similarly, <u>disjunction</u>, written

$$\vee : \Omega \times \Omega \longrightarrow \Omega \quad ,$$

is the characteristic map of

$$(\Omega \times 1) \cup (1 \times \Omega) \rightarrowtail \Omega \times \Omega \quad .$$

Now we can define the order relation $\Omega_0 \rightarrowtail \Omega \times \Omega$ as an equalizer

$$\Omega_0 \rightarrowtail \Omega \times \Omega \underset{\wedge}{\overset{\pi_1}{\rightrightarrows}} \Omega \quad ,$$

and its characteristic map gives the operation

$$\Longrightarrow : \Omega \times \Omega \longrightarrow \Omega \quad ,$$

which we call implication. With respect to these operations, one proves that Ω is a Heyting algebra object in $\underline{E}$. For category theorists, this means that Ω is a trivial category object (by means of the ordering defined above) which is finitely complete and cocomplete and cartesian closed ($\Longrightarrow$ is the exponentiation for the product $\wedge$). Continuing in this vein we can define negation, written

$$\neg : \Omega \longrightarrow \Omega \quad ,$$

as the characteristic map of $1 \xrightarrow{\text{false}} \Omega$. Underscoring the intuitionistic character of topos we have the fact that

$$\neg\neg : \Omega \longrightarrow \Omega$$

is rarely the identity. For example, if T is a T_0-space, then in Sheaves (T), $\neg\neg = \text{id}$ iff T is discrete. However, we will certainly want our set theory to be classical, and this forms the content of the next axiom. From among several possible ways of stating this we choose the following: $1 \xrightarrow{\text{false}} \Omega$ and $1 \xrightarrow{\text{true}} \Omega$ Provide a map $1 + 1 \longrightarrow \Omega$, and we require

Axiom 5. $1 + 1 \longrightarrow \Omega$ is an isomorphism.

Other conditions we might have used are: Ω is a Boolean algebra, $\neg\neg = \text{id}$, subobjects have complements, etc. We call a topos Boolean if it satisfies Axiom 5.

We shall further require the axiom of choice, which we state as

Axiom 6. Epimorphisms split.

That is, for every epimorphism $q : X \longrightarrow\!\!\!\!\!> Q$ (such arrows always denote epimorphisms) there exists a map $s : Q \longrightarrow X$ such that $qs = \text{id}$. s is called a section for q .

Axioms 1 - 6 describe Boolean set theory - with the axiom of choice-and perhaps a remark should be made about that. Namely, for X in an arbitrary topos $\underline{E}$, factoring the canonical map $X \longrightarrow 1$

yields

$$\begin{array}{ccc} X & \longrightarrow & 1 \\ & \searrow\!\!\!\!\searrow \quad \nearrow & \\ & \sigma(X) & \end{array}$$

where we call $\sigma(X)$ the <u>support</u> of X . If the epi part of the above factorization always splits, we say that <u>supports split</u> in $\underline{E}$. Now if $\underline{E}$ is Boolean, and supports split, then the subobjects of 1 form a system of generators for $\underline{E}$, which is a kind of extensionality condition . Thus, even when investigating Boolean set theory without AC one should still require that supports split.

Finally, although we have required Ω to be $1 + 1$, it can still be very large — in fact its global sections can be an arbitrary complete Boolean algebra. Hence, to pin down the category of sets as 2-valued set theory, we require

<u>Axiom 7</u>. If $U \rightarrowtail 1$, then $U \simeq 0$ or $U \simeq 1$.

Before going on, it is probably worthwhile to make a few remarks on this system, which we might call CS — for the category of sets. As indicated previously, Axioms 1 - 7, as a group, are equivalent to the axioms given in [3], though the emphasis here is placed quite differently. Roughly speaking, the point of view here is that the category of sets is merely one among many topos, and one should learn not to specialize too soon. In [3], for example, much more weight was placed on the axiom of choice, whose independence could be shown by considering the category of partially ordered sets, this being a model of all the axioms except AC. Here this will never work, since the category of partially ordered sets is far from being even a topos, which is as it should be.

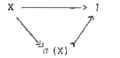

The extent to which CS describes, say, the category of sets built from ZF will be touched on at the end of the paper. Suffice to say here that, though the axiom of replacement is lacking, one seems to be able to develop most of the mathematics one would like to

in CS . We might say a few words about uniqueness, however. Namely, suppose $\underline{S}$ is a model for CS. Then we can prove that topos defined over $\underline{S}$, in which the subjobjects of 1 generate , are in $1-1$ correspondence with complete Heyting algebras in $\underline{S}$. In particular, then, any other model of CS defined over $\underline{S}$ is equivalent to $\underline{S}$. This is the precise form of the statement that if you add the non-elementary axiom of completeness, then you have characterized set theory.

Now, for a moment, let us discuss the Continuum Hypothesis. This is a categorical question which we can phrase as follows: Does there exist an X such that

$$\omega \rightarrowtail X \rightarrowtail 2^{\omega}$$

properly? That is, $X \not\cong \omega$ and $X \not\cong 2^{\omega}$. What we shall do is to start with a model $\underline{S}$ of CS and construct another model $\underline{S}'$ in which such an X exists. Very briefly, the procedure is this: By considering functors on partially ordered sets in $\underline{S}$, we can find many models for Axioms 1-4 . By passing to sheaves in these "presheaf" categories, we can pick up Axioms 5 and 6. By choosing a particular partially ordered set we can negate CH in the internal topos logic, which is much stronger than merely negating it in the external form above (though equivalent in the 2-valued case). Now, collapsing along a morphism that preserves the topos structure we can find a 2-valued model in which CH is false. Let us begin with the general construction of models for Boolean set theory. In our fixed model $\underline{S}$ of CS, we will use an ε relation defined as in [3]. Although our use of it here is somewhat informal, the dilegent reader should have no trouble supplying any desired details.

Call the objects of $\underline{S}$ sets, and let $\mathbb{P}$ be a partially ordered set. Then $\mathbb{P}$ is a trivial category via its ordering — i.e., $p \longrightarrow q$ iff $p \leq q$ — and we can form the category $\underline{S}^{\mathbb{P}}$ of covariant $\underline{S}$-valued functors on $\mathbb{P}$. These functors should be thought of as sets parametrized by $\mathbb{P}$. That is, $X \in \underline{S}^{\mathbb{P}}$ is given by specifying its value at $p \in \mathbb{P}$; $X(p) \in \underline{S}$; and requiring for $p \leq q$ that $X(p) \longrightarrow X(q)$ in a transitive way. $\mathbb{P}$ itself is embedded in $\underline{S}^{\mathbb{P}}$ by the Yoneda functor which sends $p \in \mathbb{P}$ into the representable functor defined by p (also written simply as p) — i.e., $p \in \underline{S}^{\mathbb{P}}$ is given by

$$p(q) = \begin{cases} 1 & \text{if } p \leq q \\ 0 & \text{otherwise} \end{cases} .$$

Since $\underline{S}^{\mathbb{P}}$ is a special case of the construction $\underline{S}^{\underline{C}}$ discussed earlier, $\underline{S}^{\mathbb{P}}$ is a topos. Moreover, $\underline{S}^{\mathbb{P}}$ is defined over $\underline{S}$ by means of the adjoint pair

$$\underline{S}^{\mathbb{P}} \underset{(1,-)}{\overset{\Delta}{\rightleftarrows}} \underline{S}$$

where $(1,-)$ at $X \in \underline{S}^{\mathbb{P}}$ is the set of maps — in $\underline{S}^{\mathbb{P}}$ — from 1 to X , and $\Delta S(p) = S$ for all $p \in \mathbb{P}$. Therefore, Axiom 4 holds in $\underline{S}^{\mathbb{P}}$. Recall that $\Omega \in \underline{S}^{\mathbb{P}}$ is the functor whose value at $p \in \mathbb{P}$ is the collection of subobjects — in $\underline{S}^{\mathbb{P}}$ — of the representable functor p . If $p \leq q$, then $\Omega(p) \longrightarrow \Omega(q)$ is given by pulling back the subobject $R \rightarrowtail p$ along the map $q \longrightarrow p$. Now such a subobject is completely determined by specifying those $q \geq p$ at which R takes the value 1, so we will identify subobjects of p with filters of elements of $\mathbb{P}$ that are $\geq p$ — i.e., sets of elements R of $\mathbb{P}$ that are $\geq p$ and have the property that if $q \in R$ and $q' \geq q$, then $q' \in R$. The map $1 \xrightarrow{\text{true}} \Omega$ picks out, for every $p \in \mathbb{P}$, the filter of all elements $\geq p$, i.e., p itself, and $1 \xrightarrow{\text{false}} \Omega$ picks out the empty filter for each p.

Now, although $\underline{S}^{\mathbb{P}}$ is a topos, it will not be Boolean unless $\mathbb{P}$ is discrete, so we must go further in order to obtain a Boolean topos. Since the deviation of $\neg\neg : \Omega \longrightarrow \Omega$ from the identity measures the failure of $\underline{S}^{\mathbb{P}}$ to be Boolean, let us calculate this map explicitly. Since $\neg:\Omega \longrightarrow \Omega$ is the characteristic map of $1 \xrightarrow{\text{false}} \Omega$, it follows that if $X' \rightarrowtail X$ has characteristic map $\phi:X \longrightarrow \Omega$, then $\neg X' \rightarrowtail X$ — the subobject corresponding to $\neg\phi$ — is the subfunctor given by

$$\neg X'(p) = \{x \in X(p) \mid \phi(p)(x) = 0\} \quad .$$

If $q \geq p$, then $X(p) \longrightarrow X(q)$, and writing x_q for the value of this map on $x \in X(p)$, we have

$$\neg X'(p) = \{x \in X(p) \mid \forall\, q \geq p,\ x_q \notin X'(q)\} \quad .$$

To illustrate the correspondence of subobjects with characteristic maps, let us prove this. So suppose $x \in X(p)$ is such that $\forall q \geq p,\ x_q \notin X'(q)$ but $\phi(p)(x) = R \rightarrowtail p$ is not 0. Then for some $q \geq p$, $R(q) = 1$, i.e.,

$$\begin{array}{ccc} q & \longrightarrow & R \\ \| & & \downarrow \\ q & \longrightarrow & p \end{array}$$

is a pullback. But then, since the diagram

$$\begin{array}{ccccc} X'(p) & \rightarrowtail & X(p) & \xrightarrow{\phi(p)} & \Omega(p) \\ \downarrow & & \downarrow & & \downarrow \\ X'(q) & \rightarrowtail & X(q) & \xrightarrow{\phi(q)} & \Omega(q) \end{array}$$

commutes, we have $\phi(q)(x_q) = q$, or $x_q \in X'(q)$. The reverse implication is similar. Applying $\neg$ we see that

$$\neg\neg X'(p) = \{x \in X(p) \mid \forall q \geq p \exists r \geq q \text{ with } x_r \in X'(r)\} .$$

Clearly, $X' \rightarrowtail \neg\neg X'$. One calls $X' \rightarrowtail X$ <u>dense</u> if $\neg\neg X' = X$, and <u>closed</u> if $\neg\neg X' = X'$. By the above claculations, $X' \rightarrowtail X$ is dense iff for any $p \in \mathbb{P}$ and $x \in X(p)$ there exists $r \geq p$ with $x_r \in X'(r)$, and closed iff, given $x \in X(p)$ such that $\forall q \geq p \exists r \geq q$ with $x_r \in X'(r)$, it follows that $x \in X'(p)$.

If we identify subobjects $R \rightarrowtail p$ with filters of elements of $\mathbb{P} \geq p$, and use the Yoneda correspondence between maps $p \longrightarrow \Omega$ and elements of $\Omega(p)$ we find that if $R \rightarrowtail p$, then

$$\neg R = \{q \geq p \mid \forall q' \geq q,\ q' \notin R\}$$

and

$$\neg\neg R = \{q \geq p \mid \forall q' \geq q \exists r \geq q' \text{ with } r \in R\} .$$

Thus, $R \rightarrowtail p$ is dense iff $\forall q \geq p \exists r \geq q$ such that $r \in R$, and closed iff, given $q \geq p$ such that $\forall q' \geq q \exists r \geq q$ with $r \in R$, then $q \in R$. Note that if $\mathbb{P}$ has a maximal element, then any nonempty crible is dense.

In a moment we shall show that the collection of dense subobjects of p, for each $p \in \mathbb{P}$, forms a Grothendieck topology on $\mathbb{P}$ called the $\neg\neg$-<u>topology</u>. Before doing this, however, let us remark that in the general treatment it is $\neg\neg$ itself that is the topology, and it is present in any topos. In a topos of the form $\underline{S}^{\underline{C}}$, a topology given by an endomorphism of Ω satisfying certain axioms is equivalent to a Grothendieck topology on $\underline{C}$, but the former makes sense in any topos and is idependent of the notion of "site." For now, however, let us simply verify the Grothendieck-

Verdier axioms [2] for a topology:

(i) For $p \in \mathbb{P}$, $p \xrightarrow{id} p$ is dense (clear)

(ii) If $q \geq p$ — i.e., $q \longrightarrow p$ in $\underline{S}^{\mathbb{P}}$ — and $R \rightarrowtail p$ is dense, so is $R|q \rightarrowtail q$ (clear).

(iii) Suppose $R' \rightarrowtail p$, and $R \rightarrowtail p$ is a dense crible with the property that $\forall q \in R$, $R'|q \rightarrowtail q$ is dense. We must show $R' \rightarrowtail p$ is dense. So, suppose $q \geq p$. Then, since $\exists r \geq q$ such that $r \in R$, $R'|r \rightarrowtail r$ is dense. In particular, $\exists r' \geq r$ with $r' \in R'$ and so $R' \rightarrowtail p$ is dense.

Now, having a topology we may consider separated objects, sheaves, etc. Again we remark that this process of passing to sheaves can — and should — be carried out in an arbitrary topos. Here, however, we will say $X \in \underline{S}^{\mathbb{P}}$ is <u>separated</u> iff for any $p \in \mathbb{P}$ and any dense $R \rightarrowtail p$, the canonical map

$$X(p) \longrightarrow \varprojlim_{q \in R} X(q)$$

is monic. X is called a $\neg\neg$-<u>sheaf</u> if the above map is an isomorphism. As reinforcement for one's topological intiution, one readily checks that X is separated iff the diagonal $X \rightarrowtail X \times X$ is closed. In the same vein, X is a sheaf iff it is separated and <u>absolutely closed</u>, i.e., closed in any separated object containing it. Notice, for later reference, that if $S \in \underline{S}$, then ΔS — the constant presheaf at S — is certainly separated, but need not be a sheaf.

Passing to sheaves, let

$$\mathrm{Sh}_{\neg\neg}(\mathbb{P}) \xrightarrow[i]{} \underline{S}^{\mathbb{P}}$$

denote the inclusion of the full subcategory of sheaves for the $\neg\neg$-topology. From [2], or better, [4], we take the following facts: There is an associated sheaf functor

$$a: \underline{S}^{\mathbb{P}} \longrightarrow \mathrm{Sh}_{\neg\neg}(\mathbb{P})$$

which is left adjoint to i and preserves finite limits. Moreover, if $X \in \underline{S}^{\mathbb{P}}$ is separated, then the unit $X \longrightarrow ia(X)$ is monic and dense. Composing the adjoint pairs

$$\mathrm{Sh}_{\neg\neg}(\mathbb{P}) \underset{i}{\overset{a}{\rightleftarrows}} \underline{S}^{\mathbb{P}} \underset{(1,-)}{\overset{\Delta}{\rightleftarrows}} \underline{S} \quad ,$$

we obtain a single pair

$$\mathrm{Sh}_{\neg\neg}(\mathbb{P}) \underset{\Gamma}{\overset{\wedge}{\rightleftarrows}} \underline{S} \quad .$$

Thus, if X is a $\neg\neg$-sheaf $\Gamma(X)$ = set of global sections of X , and if $S \in \underline{S}$, then $\hat{S}$ is the sheaf associated to the constant presheaf at S.

Here we pause for a moment to prove a technical lemma that will be useful later on. Namely, suppose $S \in \underline{S}$. Then we have

<u>Lemma 1</u>. For any section $p \xrightarrow{x} \hat{S}$, there is an element $s \in S$ and an $r \geq p$ such that

$$\begin{array}{ccccc} & & p & & \\ & \nearrow & & \searrow{\scriptstyle x} & \\ r & & & & \hat{S} \\ & \searrow & & \nearrow{\scriptstyle \hat{s}} & \\ & & 1 & & \end{array}$$

commutes.

<u>Proof</u>: As above, we have a dense monomorphism $\Delta S \rightarrowtail \hat{S}$. Thus, as was established previously, for any $p \in \mathbb{P}$ and $x \in \hat{S}(p)$, $\exists r \geq p$ such that $x_r \in \Delta S(r)$. By the Yoneda lemma, this is the same as saying that for any map $p \xrightarrow{x} \hat{S}$, $\exists r \geq p$ such that $r \longrightarrow p \xrightarrow{x} \hat{S}$ factors through $\Delta S \rightarrowtail S$ — i.e., we have a diagram

$$\begin{array}{ccc} r & \longrightarrow & p \\ \vdots & & \downarrow{\scriptstyle x} \\ \Delta S & \rightarrowtail & \hat{S} \end{array} .$$

$\Delta S(r) = S$, so identify $r \longrightarrow \Delta S$ with an element $s \in S$. Thinking of s as a map $1 \xrightarrow{s} S$ we have by naturality a commutative diagram

$$\begin{array}{ccc} \Delta 1 & \xrightarrow{\sim} & \hat{1} \\ {\scriptstyle \Delta s}\downarrow & & \downarrow{\scriptstyle \hat{s}} \\ \Delta S & \rightarrowtail & \hat{S} \end{array} .$$

Now composing on the right with the canonical map $r \longrightarrow \Delta 1$ gives the result.

Notice that the assertion of the lemma is the analogue of the statement for sheaves on a topological space that sections of a constant sheaf are locally constant functions into the fibre.

Our first theorem is

<u>Theorem 1</u>. $Sh_{\neg\neg}(\mathbb{P})$ is a model of Boolean set theory.

<u>Proof</u>: We must establish Axioms 1-6 in $Sh_{\neg\neg}(\mathbb{P})$. Well, Axiom 1 is trivial, and Axiom 2 holds since it is easy to verify that if $X \in Sh_{\neg\neg}(\mathbb{P})$ and Y is any presheaf, then X^Y is a sheaf. Also,

Axiom 4 is trivial, since $Sh_{\neg\neg}(\mathbb{P})$ is defined over $\underline{S}$.

What about the subobject classifier of Axiom 3? First of all, in any topos — in particular $\underline{S}^{\mathbb{P}}$ — Ω is injective, so if we define $\Omega_{\neg\neg}$ as an equalizer

$$\Omega_{\neg\neg} \rightarrowtail \Omega \underset{\mathrm{id}}{\overset{\neg\neg}{\rightrightarrows}} \Omega \quad ,$$

it follows that $\Omega_{\neg\neg}$ is also injective. To prove that $\Omega_{\neg\neg}$ is a sheaf, then, we need only show that it is separated. So, suppose $R \rightarrowtail p$ and $R' \rightarrowtail p$ are closed subobjects of p that have the same restriction to each element q of some dense $R_0 \rightarrowtail p$. If $R \neq R'$, then there is a $q \in R$, say, such that $q \notin R'$. R_0 is dense, so that $\forall q' \geq q \exists r \in R_0$ with $r \geq q'$. $r \geq q'$, and $q \in R$, so $r \in R$. But $R|r = R'|r$, hence $r \in R'|r$, i.e., $r \in R'$. Since $R' \rightarrowtail p$ is closed, $q \in R'$, which is a contradiction. Thus $R = R'$ and $\Omega_{\neg\neg}$ is a sheaf. Since $\Omega_{\neg\neg}$ classifies closed subobjects by definition, it follows that if $X' \rightarrowtail X$ is closed and X is a sheaf, so is X' (the pullback of a sheaf is a sheaf). On the other hand, if X' is a sheaf then $X' \rightarrowtail X$ is closed since X is separated and X' is absolutely closed. All of this adds up to the fact that the Ω for $Sh_{\neg\neg}(\mathbb{P})$ is precisely $\Omega_{\neg\neg}$. Moreover, from a standard Heyting algebra argument we can deduce that $\Omega_{\neg\neg}$ is a Boolean algebra object. Thus $Sh_{\neg\neg}(\mathbb{P})$ is a Boolean topos — i.e., Axiom 5 holds.

Finally, we establish Axiom 6 in $Sh_{\neg\neg}(\mathbb{P})$ by using the existence of complements for arbitrary subobjects, and Zorn's lemma in $\underline{S}$. To start, let $X \overset{f}{\twoheadrightarrow} Y$ be an epimorphism in $Sh_{\neg\neg}(\mathbb{P})$. We claim that if $Y \neq 0$, then there exists a partial section of f with non-zero domain — i.e., a commutative diagram

$$\begin{array}{ccc} & & X \\ & \overset{s}{\nearrow} & \downarrow f \\ Y_0 & \rightarrowtail & Y \end{array}$$

with $Y_0 \neq 0$. In fact, this is true for an arbitrary f with $X \neq 0$. Namely, it is easy to see that the subobjects of 1 form a system of generators in $\neg\neg$-Sh($\mathbb{P}$) . Thus, if $X \neq 0$ there is a map $U \longrightarrow X$ with U a non-zero subobject of 1 (otherwise $0 \rightarrowtail X$ is an isomorphism). Now just take $Y_0 \rightarrowtail Y$ to be the composite $U \longrightarrow X \xrightarrow{t} Y$. If f is epic, though, it must actually split globally. To see this, consider the partially ordered set (in $\underline{S}$) of all such partial sections. It is non-empty clearly, and inductive since $Sh_{\neg\neg}(\mathbb{P})$ is a topos. By Axiom 6 in $\underline{S}$, let

$$\begin{array}{ccc} & & X \\ & \overset{s}{\nearrow} & \downarrow f \\ Y_0 & \rightarrowtail & Y \end{array}$$

be a maximal element. Then $Y_0 \neq Y$ iff $Y - Y_0 \neq 0$. Forming the pullback

$$\begin{array}{ccc} X' & \longrightarrow & X \\ f' \downarrow & & \downarrow f \\ Y - Y_0 & \rightarrowtail & Y \end{array} ,$$

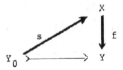

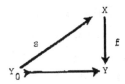

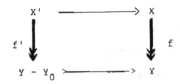

f' is epic, so if $Y - Y_0 \neq 0$ we can find a partial section of f' with non-zero domain, thus properly extending s and contradicting its maximality. Thus $Y_0 = Y$, f splits, and $Sh_{\neg\neg}(\mathbb{P})$ satisfies the axiom of choice.

We discuss next how to retrieve a 2-valued model from $Sh_{\neg\neg}(\mathbb{P})$. First, we remark that $B = \Gamma(\Omega_{\neg\neg})$ is a complete Boolean algebra in $\underline{S}$. Next, using Zorn's lemma, choose a Boolean homomorphism $h: B \longrightarrow 2$. If $X' \rightarrowtail X$ is a monomorphism in $Sh_{\neg\neg}(P)$, let $\Pi_X X' \rightarrowtail 1$ denote the result of applying to $X' \rightarrowtail X$ the right adjoint to pulling back along $X \longrightarrow 1$. Let $\phi_{X'}: 1 \longrightarrow \Omega_{\neg\neg}$ be its characteristic map. Now put

$$\Sigma = \{X' \rightarrowtail X \mid h(\phi_{X'}) = \text{true}\} \quad ,$$

write $\underline{S}(\mathbb{P},h)$ for the category of fractions $Sh_{\neg\neg}(\mathbb{P})[\Sigma^{-1}]$, and let

$$P: Sh_{\neg\neg}(\mathbb{P}) \longrightarrow \underline{S}(\mathbb{P},h)$$

denote the canonical projection. Then we have

Theorem 2. $\underline{S}(\mathbb{P},h)$ is a model of set theory. Moreover, P preserves finite limits, finite colimits, exponentiation, Ω, and ω.

Proof: We sketch only the particular aspects of the proof, since the formation of $\underline{S}(\mathbb{P},h)$ is a special case of a general construction discussed in [4]. That is, we cite [4] for the proof that $\underline{S}(\mathbb{P},h)$ is a topos and P preserves the topos structure. If we call such a functor a logical morphism of topos, then, very roughly, P is a logical morphism because Σ has a calculus of right fractions and is closed under exponentiation and its saturation $\bar{\Sigma}$ has a calculus of left fractions. In any case, it follows that P preserves the statement

$$1 + 1 \xrightarrow[\sim]{} \Omega \quad ,$$

so that $\underline{S}(\mathbb{P},h)$ is also Boolean. What about Axioms 6 and 7 for $\underline{S}(\mathbb{P},h)$? Well, recall from [1], that a map $f:X \longrightarrow Y$ in $\underline{S}(\mathbb{P},h)$ can be represented in the form

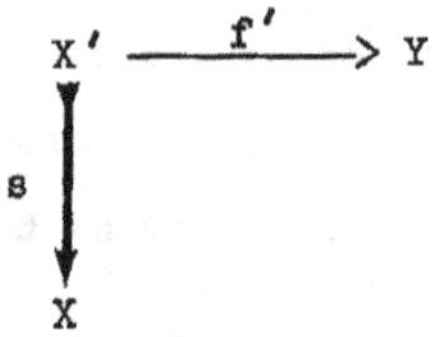

where $s \in \Sigma$. Now factor f' in $Sh_{\neg\neg}(\mathbb{P})$, giving a diagram

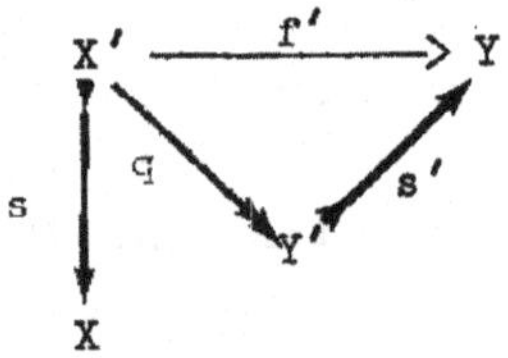

Since P preserves both epimorphisms and monomorphisms (and $\underline{S}(\mathbb{P},h)$ is a topos), it follows that if f is epic, $s' \in \Sigma$, in which case f splits in $\underline{S}(\mathbb{P},h)$ since q splits in $Sh_{\neg\neg}(\mathbb{P})$. Thus the axiom of choice holds in $\underline{S}(\mathbb{P},h)$. Also, if f is monic then $q \in \bar{\Sigma}$ — the saturation of Σ. In particular, suppose f is of the form $U \rightarrowtail 1$ in $\underline{S}(\mathbb{P},h)$. Then the above factorization looks like

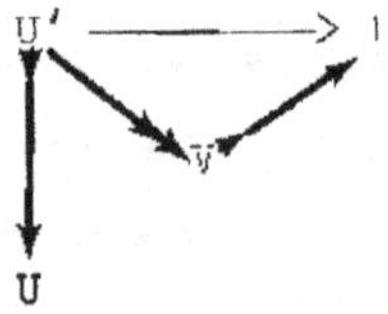

with $U' \twoheadrightarrow V$ in $\bar{\Sigma}$. Let $\varphi_V : 1 \longrightarrow \Omega_{\neg\neg}$ be the characteristic map of $V \rightarrowtail 1$ in $Sh_{\neg\neg}(\mathbb{P})$. Then either $h(\varphi_V) = \text{true}$ or $h(\varphi_V) = \text{false}$. If $h(\varphi_V) = \text{true}$ then $V \rightarrowtail 1$ is in Σ, so $U \rightarrowtail 1$ is an isomorphism in $\underline{S}(\mathbb{P},h)$. If $h(\varphi_V) = \text{false}$, then

$h(\neg\varphi_V) = true$ and $\neg V \rightarrowtail 1$ is in Σ. But

$$\begin{array}{ccc} 0 & \rightarrowtail & \neg V \\ \downarrow & & \downarrow \\ V & \rightarrowtail & 1 \end{array}$$

is a pullback in $Sh_{\neg\neg}(\mathbb{P})$, so $0 \rightarrowtail V$ is in Σ. But then U is isomorphic to 0 in $\underline{S}(\mathbb{P},h)$, and $\underline{S}(\mathbb{P},h)$ satisfies Axiom 7.

Provided with Theorems 1 and 2, we can give a more precise idea of how it is one shows CH does not follow from Axioms 1-7. First, given X and Y in any topos $\underline{E}$, we can define the <u>internal</u> object of epimorphisms $Epi(X,Y)$. To do this, start by defining the map "image": $Y^X \longrightarrow \Omega^Y$. To give such a map we need a map $Y^X \times Y \longrightarrow \Omega$ or, equivalently, a subobject of $Y^X \times Y$. For this, take the image $S \rightarrowtail Y^X \times Y$ of the map

$$\begin{array}{ccc} Y^X \times X & \xrightarrow{\langle \pi_1, ev \rangle} & Y^X \times Y \\ & \searrow \quad \nearrow & \\ & S & \end{array} ,$$

where ev is the evalution map. It is easy to see that if $f:X \longrightarrow Y$, then the composite

$$1 \xrightarrow{\bar{f}} Y^X \longrightarrow \Omega^Y$$

is the transpose of the characteristic map of the image of f. Letting $true_Y$ be the map $Y \longrightarrow 1 \xrightarrow{true} \Omega$, we define $Epi(X,Y)$ as a pullback

$$\begin{array}{ccc} Epi(X,Y) & \rightarrowtail & Y^X \\ \downarrow & & \downarrow im \\ 1 & \xrightarrow{\overline{true_Y}} & \Omega^Y \end{array}$$

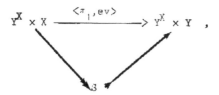

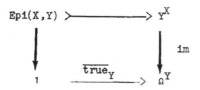

Thus, the global sections of $Epi(X,Y)$ correspond to the actual epimorphisms $X \longrightarrow Y$, but there will also be many sections over other objects. Now the procedure is simple. Namely, we shall choose $\mathbb{P}$ in such a way that in $Sh_{\neg\neg}(\mathbb{P})$ there will exist an X with the property (remember $\Omega_{\neg\neg} \cong 1 + 1 = 2$)

$$\omega \rightarrowtail X \rightarrowtail 2^{\omega}$$

but $Epi(\omega,X) = Epi(X,2^{\omega}) = 0$. Once we do this we are finished, for $P: Sh_{\neg\neg}(\mathbb{P}) \longrightarrow \underline{S}(\mathbb{P},h)$ preserves the topos structure, thus it also preserves such an internal negation of CH.

Now, to pick $\mathbb{P}$ first choose an $I \in \underline{S}$ such that $2^{\omega} \rightarrowtail I$ but $2^{\omega} \not\twoheadrightarrow I$ — this can be done by Cantor's diagonal argument. Then let $\mathbb{P}$ be the collection of partial maps from $I \times \omega$ to 2 with finite domain — i.e., a $p \in \mathbb{P}$ is a partial map

$$\begin{array}{ccc} F & \longrightarrow & 2 \\ \downarrow & & \\ I \times \omega & & \end{array}$$

where F is finite and $F \longrightarrow 2$ is arbitrary. Put $p \leq q$ iff $\operatorname{dom} p \subset \operatorname{dom} q$ and $q|\operatorname{dom} p = p$.

One important reason for choosing $\mathbb{P}$ to be a set of partial maps with finite domain (and codomain) is that any such set satisfies the ω-chain condition. More precisely, let T be a finite set with $\operatorname{card}(T) = m$, and let $J \in \underline{S}$. If $\mathbb{Q}$ is the object of partial maps from J to T with finite domain, then $\mathbb{Q}$ has the following property:

Lemma 2. If $Z \rightarrowtail \mathbb{Q}$ is such that no pair q,q' in Z has a common extension in $\mathbb{Q}$ — meaning there exists no p in $\mathbb{Q}$ with $p \geq q$ and $p \geq q'$ — then $\omega \twoheadrightarrow Z$ — i.e., Z is countable.

Proof: We first reformulate the condition on Z. Namely, for $q,q' \in Z$ there exists no p in $\mathbb{Q}$ with $p \geq q$ and $p \geq q'$ iff there is a $j \in dom(q) \cap dom(q')$ such that $q(j) \neq q'(j)$. Now suppose that for each $q \in Z$ $card(dom(q)) = n$. Then we claim

$$card(Z) \leq n!m^n \quad ,$$

which we prove by induction on n. The statement being clearly true for $n = 1$, assume it for n and suppose $card(dom(q)) = n + 1$ for all $q \in Z$. Then, for $j \in J$ and $t \in T$, let

$$Z(j,t) = \{q \in Z \mid j \in dom(q) \text{ and } q(j) = t\} \quad .$$

Fixing some $q_0 \in Z$, we have

$$Z = \bigcup_{\substack{j \in dom(q_0) \\ t \in T}} Z(j,t) \quad .$$

For each $Z(j,t)$, form Z' by removing j from $dom(q)$ for all $q \in Z(j,t)$. Now if $q \neq q'$ in $Z(j,t)$ then there is a $j' \in dom(q) \cap dom(q')$ such that $q(j') \neq q'(j')$. Thus $j' \neq j$, so

$$card(Z(j,t)) = card(Z')$$

and Z' still satisfies the original condition on Z. But now, $card(dom(q) = n$ for all $q \in Z'$, so

$$card(Z') \leq n!m^n \quad .$$

But this gives

$$\begin{aligned} card(Z) &\leq (n+1)mn!m^n \\ &= (n+1)!m^{n+1} \quad . \end{aligned}$$

Now, of course, for an arbitrary Z satisfying the given disjointness condition, if

$$Z_n = \{q \in Z \mid card(dom(q)) = n\}$$

$Z = \bigcup_n Z_n$ so Z is countable.

We might remark that, although this result suffices for our purposes, one can obviously use the same proof to get various stronger results. For example, T could be countable, or one could prove a more general cardinality result from the start. Also, we should probably say that we call this the ω-chain condition because if one forms $Sh_{\neg\neg}(\mathbb{Q})$, then the above condition on $\mathbb{Q}$ is equivalent to the usual ω-chain condition on the Boolean algebra $B = \Gamma(\Omega_{\neg\neg})$.

Going back to our original $\mathbb{P}$, consider the adjoint pair

$$Sh_{\neg\neg}(\mathbb{P}) \underset{\Gamma}{\overset{\wedge}{\rightleftarrows}} \underline{S} \quad .$$

The importance of the ω-chain condition is that it enables us to establish one of the two basic results used in negating CH. Namely,

<u>Theorem 3</u>. If $X,Y \in \underline{S}$ and X is infinite (meaning $X \neq 0$ and $X \cong X \times \omega$) then $Epi(X,Y) = 0$ in $\underline{S}$ yields $Epi(\hat{X},\hat{Y}) = 0$ in $Sh_{\neg\neg}(\mathbb{P})$.

<u>Proof</u>: Suppose $Epi(\hat{X},\hat{Y}) \neq 0$. Then there is a map $p \longrightarrow Epi(\hat{X},\hat{Y})$ for some $p \in \mathbb{P}$. But this, it is easy to see, is the same as a map $f': p \times \hat{X} \longrightarrow \hat{Y}$ such that $\langle\pi_1, f'\rangle: p \times \hat{X} \twoheadrightarrow p \times \hat{Y}$ is epic. By Booleanness one can extend f' to $f: \hat{X} \longrightarrow \hat{Y}$ — i.e., there is a diagram

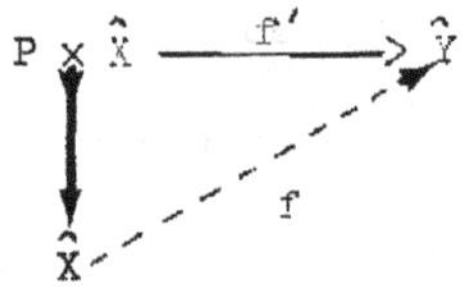

This is so because $\hat{Y}$ has global sections onto any one of which we can map the complement of $p \times \hat{X} \rightarrowtail \hat{X}$. Now $p \times f = \langle\pi_1, f'\rangle$, so $p \times f: p \times \hat{X} \twoheadrightarrow p \times \hat{Y}$ is epic. In $\underline{S}$, define $E \rightarrowtail \mathbb{P} \times X \times Y$ by

$$E = \{(p,x,y) \mid p \rightarrowtail 1 \;\underset{\hat{y}}{\overset{\hat{x}}{\longrightarrow}}\; \hat{X} \overset{f}{\longrightarrow} \hat{Y} \text{ commutes}\}$$

Let $E \longrightarrow \mathbb{P}$, $E \longrightarrow X$, and $E \longrightarrow Y$ be the maps obtained by composing $E \rightarrowtail \mathbb{P} \times X \times Y$ with the various projections. First, we claim $E \longrightarrow Y$ is epic. Well, consider $1 \overset{y}{\rightarrowtail} Y$ and

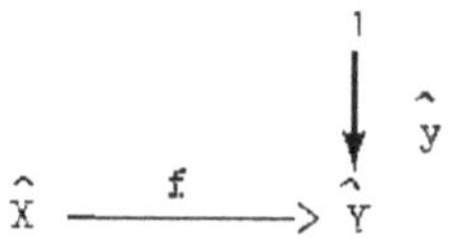

Crossing with p yields

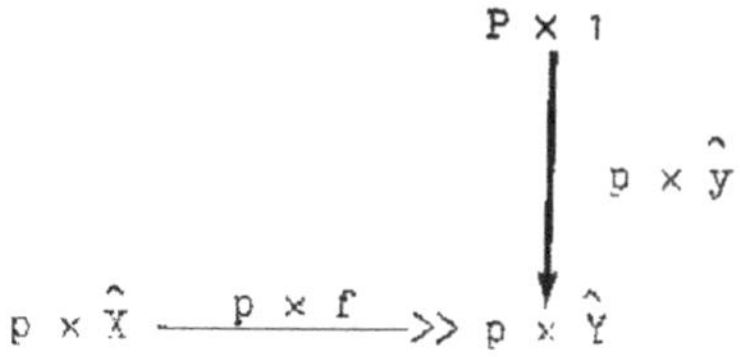

By AC in $Sh_{\neg\neg}(\mathbb{P})$, $p \times f$ splits, so there is a map $\sigma : p \times 1 \longrightarrow p \times X$ such that

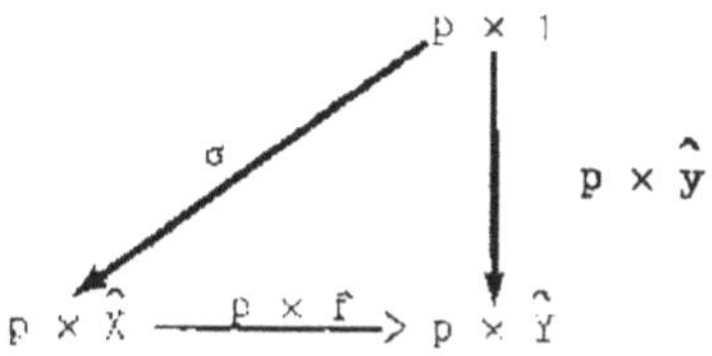

commutes. Composing with π_2 yields a diagram

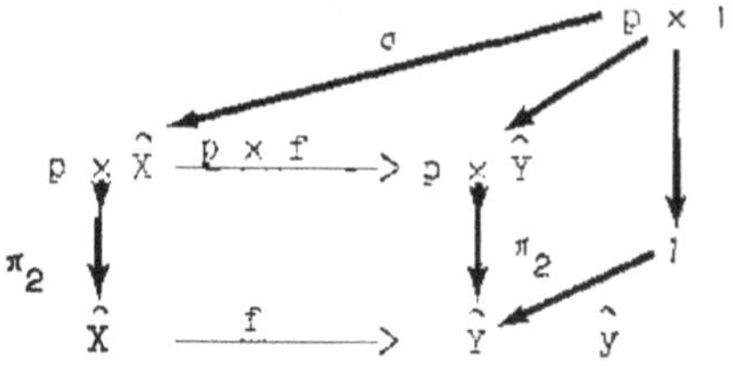

If we write σ' for $\pi_2 \circ \sigma$, then

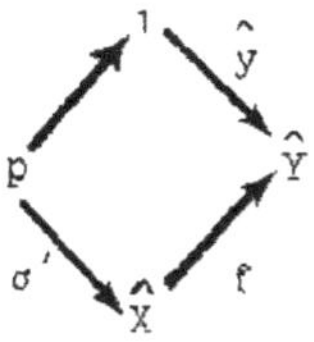

commutes. But then by Lemma 1, $\exists q \geq p$ and $x \in X$ such that

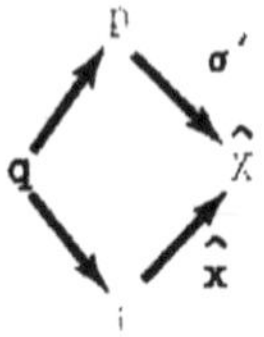

commutes. Thus the two composites in

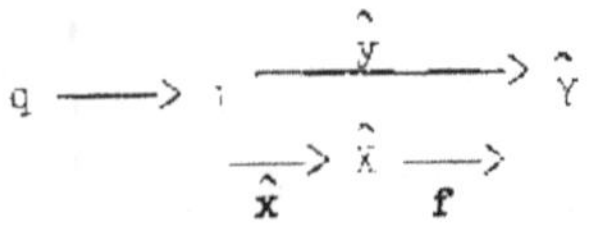

are equal, so $(q,x,y) \in E$. By AC in $\underline{S}$, let $Y \xrightarrow{s} E$ be a section for $E \longrightarrow Y$. If $1 \xrightarrow{x} X$, let Y_x be the pullback

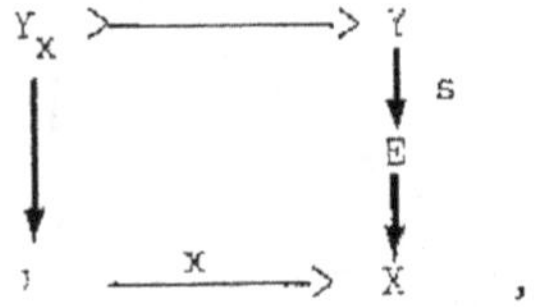

i.e.,

$$Y_x = \{y \in Y \mid s(y) = (q,x,y) \text{ for some } q\} .$$

Denote the composite $Y_x \longrightarrow Y \xrightarrow{s} E \longrightarrow \mathbb{P}$ by $p: Y_x \longrightarrow \mathbb{P}$ — i.e., if $y \in Y_x$ and $s(y) = (q,x,y)$, then $p(y) = q$. Then not only is p monic, but moreover, the condition of Lemma 2 is satisfied. That is, if $y \neq y'$ in Y_x, then there is no $p \in \mathbb{P}$ such that $q \geq p(y)$ and $q \geq p(y')$. For suppose so. Then from the diagram

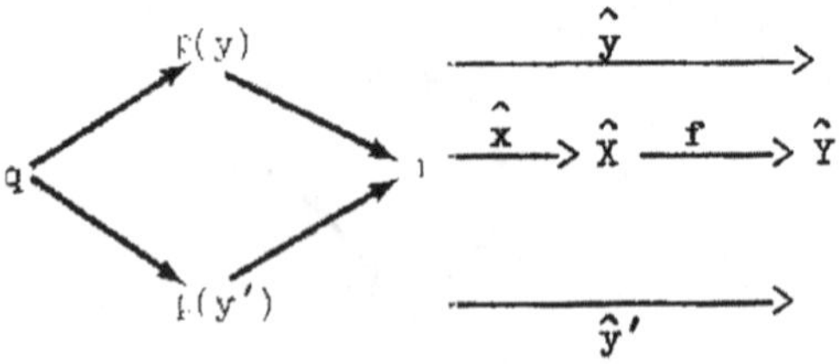

we see that

$$q \longrightarrow 1 \underset{\hat{y}'}{\overset{\hat{y}}{\rightrightarrows}} \hat{Y}$$

commutes. But in that case, the diagram

$$\begin{array}{ccccc} q & \longrightarrow & \Delta 1 & \underset{y'}{\overset{y}{\rightrightarrows}} & \Delta Y \\ & & \Big\downarrow{\scriptstyle \rho_1} & & \Big\downarrow \\ & & \hat{1} & \underset{\hat{y}'}{\overset{\hat{y}}{\rightrightarrows}} & \hat{Y} \end{array}$$

yields $y = y'$. Thus, for each $x \in X$ we have an epimorphism $\rho_x : \omega \longrightarrow\!\!\!\!\!> Y_x$ by the ω-chain condition for $\mathbb{P}$. Now define an epimorphism $\rho : X \times \omega \longrightarrow\!\!\!\!\!> Y$ by $\rho(x,n) = \rho_x(n)$. But then, since $X \simeq X \times \omega$, we have constructed an element of $\mathrm{Epi}(X,Y)$ in $\underline{S}$, and this is a contradiction.

The next step, which is the essence of forcing, is to prove

Theorem 4. In $Sh_{\neg\neg}(\mathbb{P})$ there is a monomorphism

$$\hat{I} \rightarrowtail \Omega_{\neg\neg}^{\omega} \quad .$$

If we prove this, then since $2 = 1 + 1 \simeq \Omega_{\neg\neg}$ in $Sh_{\neg\neg}(P)$, we will have been able to choose an arbitrary degree of largeness $2^{\omega} \rightarrowtail I$ in $\underline{S}$, and force 2^{ω} to be at least as large as $\hat{I}$ in $Sh_{\neg\neg}(\mathbb{P})$.

Proof: We start by giving a map

$$\varphi : \Delta I \times \Delta\omega \longrightarrow \Omega_{\neg\neg}$$

in $\underline{S}^{\mathbb{P}}$, or what is the same thing, a closed subobject

$$R \rightarrowtail \Delta I \times \Delta\omega \quad .$$

To do this, put

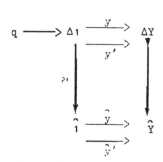

$$R(p) = \{\langle i,n\rangle \mid p\langle i,n\rangle = \text{true}\} \quad .$$

Let us check that $R \rightarrowtail \Delta I \times \Delta\omega$ is closed. Well, suppose $\langle i,n\rangle \in (\Delta I \times \Delta\omega)(p)$ is such that $\forall q \geq p \; \exists r \geq q$ with $\langle i,n\rangle_r = \langle i,n\rangle \in R(r)$ — i.e., $r\langle i,n\rangle = \text{true}$. Certainly $\langle i,n\rangle \in \text{dom}(p)$, for if not then $\exists q \geq p$ with $q\langle i,n\rangle = \text{false}$ which is impossible by the above. But then $p\langle i,n\rangle = \text{true}$, since $\exists r \geq p$ with $r\langle i,n\rangle = \text{true}$. So, let $\varphi : \Delta I \times \Delta\omega \longrightarrow \Omega_{\neg\neg}$ be the characteristic map of $R \rightarrowtail \Delta I \times \Delta\omega$. Then we claim that the transpose

$$\bar{\varphi} : \Delta I \longrightarrow \Omega_{\neg\neg}^{\Delta\omega}$$

is monic. Well, if $i,j \in I$, the the composites

$$p \underset{j}{\overset{i}{\rightrightarrows}} \Delta I \xrightarrow{\bar{\varphi}} \Omega_{\neg\neg}^{\Delta\omega}$$

are equal iff the transposes

$$p \times \Delta\omega \underset{j \times \Delta\omega}{\overset{i \times \Delta\omega}{\rightrightarrows}} \Delta I \times \Delta\omega \longrightarrow \Omega_{\neg\neg}$$

are equal, iff in

$$\begin{array}{ccccc} R_i & R_j & & & R \\ \downarrow & \downarrow & & & \downarrow \\ \multicolumn{2}{c}{p \times \Delta\omega} & \underset{j \times \Delta\omega}{\overset{i \times \Delta\omega}{\rightrightarrows}} & & \Delta I \times \Delta\omega \end{array}$$

the pullbacks R_i and R_j are equal. This is true iff $\forall q \geq p$ we have

$$\begin{array}{ccccc} R_i(q) & = \; R_j(q) & & & R(q) \\ \downarrow & \downarrow & & & \downarrow \\ \multicolumn{2}{c}{1 \times \omega} & \underset{j \times \omega}{\overset{i \times \omega}{\rightrightarrows}} & & I \times \omega \end{array}$$

But

$$R_i(q) = \{n | q\langle i,n\rangle = \text{true}\}$$

$$R_j(q) = \{n | q\langle j,n\rangle = \text{true}\} \quad ,$$

and these are not equal if $i \neq j$, since $\text{dom}(p)$ is finite — i.e., if $i \neq j$, we can find an n such that neither $\langle i,n\rangle$ nor $\langle j,n\rangle$ is in $\text{dom}(p)$. Then, however, $\exists q \geq p$ with $q\langle i,n\rangle = \text{true}$ and $q\langle j,n\rangle = \text{false}$. Now we are done, for $\Omega_{\neg\neg}$ is a sheaf, thus the dense monomorphism $\Delta\omega \rightarrowtail \hat{\omega}$ induces an isomorphism

$$\Omega_{\neg\neg}^{\hat{\omega}} \xrightarrow{\sim} \Omega_{\neg\neg}^{\Delta\omega}$$

so we have a monomorphism

$$\Delta I \rightarrowtail \Omega_{\neg\neg}^{\hat{\omega}} \quad .$$

Applying the associated sheaf functor, which preserves monomorphisms, gives a monomorphism

$$\hat{I} \overset{\psi}{\rightarrowtail} \Omega_{\neg\neg}^{\hat{\omega}}$$

which is what we want, since $\omega = \hat{\omega}$ in $Sh_{\neg\neg}(\mathbb{P})$.

To finish the argument, consider, in $\underline{S}$, the maps

$$\omega \rightarrowtail 2^{\omega} \rightarrowtail I \quad .$$

Then $\text{Epi}((\omega,2^{\omega}) = 0 = \text{Epi}(2^{\omega},I)$ — the first by Cantor's diagonal argument and the second by hypothesis. Thus in $Sh_{\neg\neg}(\mathbb{P})$ we obtain

$$\omega = \hat{\omega} \rightarrowtail \widehat{2^{\omega}} \rightarrowtail \hat{I} \overset{\psi}{\rightarrowtail} 2^{\omega} \quad ,$$

and by Theorem 3, $\text{Epi}(\omega,\widehat{2^{\omega}}) = 0 = \text{Epi}(\widehat{2^{\omega},\hat{I}})$. But I has global sections, so ψ splits (by Booleanness). Let $q: 2^{\omega} \twoheadrightarrow \hat{I}$ be a map such that $q\psi = \text{id}$. Since q is epic, it induces by composition a map

$$\text{Epi}(\widehat{2^{\omega},2^{\omega}}) \longrightarrow \text{Epi}(\widehat{2^{\omega},\hat{I}}) \quad .$$

Thus $\mathrm{Epi}(\widehat{2^{\omega}},2^{\omega}) = 0$ and $\widehat{2^{\omega}}$ is the X that negates CH.

Having now established the fact that CH does not follow from the seven axioms given for CS, it is only natural to ask what relation this result bears to that of Cohen for ZF. This, of course, amounts to asking for the relationship of CS to ZF. Very,very, briefly, the situation seems to be the following. Thinking of ZF sets as trees — i.e., as objects provided with a given (ε) relation, it becomes possible to pass back and forth between models of CS and models of set theory — in the usual sense. Starting with a model of CS, however, we can never hope to obtain in this way a model of ZF - CS, for example, is finitely axiomatisible. What is lacking is replacement. This is easy to formulate for CS, though still somewhat unclear for a general topos. We have not mentioned it here, since it plays no role in the argument — i.e., once assumed for CS, one will have to verify its presence in the various stages of the constructions, though as an axiom it will never be used in these constructions. In any case, once we add replacement to CS, the above process will yield an equivalence between models for CS and models for ZF. (This process has since been carried out by J.C. Cole [5], and W. Mitchell [6].) Thus, since CH is a categorical statement, its negation in one system will be equivalent with its negation in the other.

In closing, we might make a few remarks as to possible future uses for these sheaf theoretic methods — at least in so far as independence results in logic are concerned. Probably it is fair to say that though one can develop other logical constructions on topos that enable one to establish further classical independence results, for example AC can be handled in this way, it seems unlikely that these methods, using partially ordered sets, will yield many interesting new results in this area — largely because most of them have probably already been obtained by more standard techniques.

However, in this treatment we are able to deal with arbitrary categories of forcing conditions, not merely partially ordered sets, and this should prove to be a useful technique in model theory. For example, elementary theories themselves might prove to be interesting sites. Also, as indicated earlier, most topos are non-classical — in that Ω is not Boolean — and one can make use of this instead of discarding it by passing to $\neg\neg$ -sheaves. For example, it seems that the topological interpretation of intuitionism can be thought of simply as mathematics done in Sheaves (T) where T is a topological space. Many independence results in intuitionistic algebra and analysis should be provable by topos methods, though only the surface has been scratched to date.

REFERENCES

[1] P. Gabriel and M. Zisman, Calculus of Fractions and Homotopy Theory, Springer-Verlag, Berlin, Heidelberg, New York, 1967.

[2] A. Grothendieck and J. L. Verdier, Cohomologie étale des schémas, Séminaire de Géometrie Algébrique de l'Institut des Hautes Etudes Scientifiques, 1963-64.

[3] F.W. Lawvere, An elementary theory of the category of sets, mimeographed notes, University of Chicago.

[4] M. Tierney, Axiomatic sheaf theory, to appear in Bull. Amer. Math. Soc.

[5] J.C. Cole, Categories of sets and models of set theory, Aarhus preprint series 1970/71 no. 52.

[6] W. Mitchell, Boolean topoi and the theory of sets, mimeographed notes University of Chicago.

[7] F.W. Lawvere, Quantifiers and sheaves, Proceedings of the International Congress of Mathematicians 1970.

CLASSIFYING TOPOS

by

Jean Giraud

The basic facts about the classifying topos of a stack of groupoids were first stated in [3] and are exposed in detail in [4] Ch. VIII. This construction is useful in cohomology theory and has been introduced independently by D. Mumford to study the moduli of elliptic curves [7]. Algebraic stacks of groupoids are used in algebraic geometry cf. [1]. Here a simpler and more general approach allows us to treat the case of a stack whose fibers are not supposed to be groupoids. As a by-product we get the existence of fibered products in the bicategory of topos. This result was first announced by M. Hakim several years ago but was never published. I suspect that any written proof would have to deal with rather subtle technical difficulties about finite limits which are overcome here by the results of §1 .

If $\underline{S}$ is a site we use the work stack for the french champ [4] and prestack for prechamp (a prestack is merely a fibered category over the underlying category of the site) and split stack for champ scindé. Up to equivalence a split stack can be viewed as a sheaf of categories over $\underline{S}$ (or a category-object of the corresponding topos) satisfying some extra condition namely the patching of objects. As usual we choose and fix a universe $\underline{U}$. For clarity it should be recalled that a $\underline{U}$-topos is a special case of $\underline{U}$-site [5] and that any category can be viewed as a site such that any presheaf is a sheaf and any prestack is a stack.

§1. Left exact stacks.

A category is left exact if it admits finite limits. A functor $f:A \longrightarrow B$ between left exact categories A and B is left exact if it preserves finite limits. A site is said to be left exact if the underlying category is so. A stack C over a site $\underline{S}$ is said to be left exact if its fibers are left exact and if for any map $f:T \to S$ in $\underline{S}$ the inverse image functor induced by f between the fibers of C is left exact.

Lemma 1.1. A stack C over a left exact site $\underline{S}$ is left exact if and only if the underlying category and the structural functor $p:C \to \underline{S}$ are left exact.

The proof rests on the fact that a commutative square of C whose projection is cartesian in $\underline{S}$ is cartesian as soon as two opposite sides are $\underline{S}$-cartesian.

Lemma 1.2. A morphism $m:A \to B$ between two left exact stacks is left exact if and only if for any $S \in |\underline{S}|$ (1) the functor $m_S:A_S \to B_S$ induced by m between the fibers at S is left exact.

Proposition 1.3. Let $f:\underline{S}' \longrightarrow \underline{S}$ be a morphism between two sites (e.g. two topos) . Then the direct image (resp. inverse image) of a left exact stack and of a left exact morphism of stacks over $\underline{S}'$ (resp. $\underline{S}$) is left exact.

1.3.1. The direct image of a stack being nothing but pull-back along the underlying functor $f^*:\underline{S} \longrightarrow \underline{S}'$ of f , preserves the fibers, hence the left exactness. To treat the case of the inverse image by f of a stack over $\underline{S}$ we will use the following caracterisation of left-exactness.

1.3.2. First let I be a finite category. For any stack F over $\underline{S}$ let F^I be the prestack whose fiber at $S \in |\underline{S}|$ is the category of functors from I to the fiber F_S . One checks easily that this is a stack provided with a morphism of stacks (constant diagrams)

---

(1) The set of objects of a category C is denoted by $|C|$.

(1) $$cF : F \longrightarrow F^I \quad .$$

Furthermore F *is left exact if and only if for any finite category* I cF *admits a right adjoint in the bicategory of stacks*. The if part is obvious since such an adjoint λ induces an adjoint to each functor cF_S , $S \in |\underline{S}|$, induced by cF on the fibers at S and since λ is cartesian. The only if part is no more difficult than (1.2). Since the property of having a right adjoint is preserved by morphisms of bicategories and since the inverse image of stacks is such a morphism [4] p.88, it remains to show the following.

Lemma 1.3.3. One has a natural equivalence $e : f^*(F^I) \longrightarrow f^*(F)^I$ such that $e . f^*(cF) = cf^*(F)$.

According to [4] p.88, the inverse image of a stack F is given by the formula

(1) $$f^*(F) = Af^{-1}(LF)$$

where LF is the free split stack associated to F [4] p.39, where f^{-1} denotes the inverse image of LF as a category-object of the topos $\tilde{\underline{S}}$ and where A stands for "associated stack". Since there is a natural equivalence $F \longrightarrow LF$ and L is a morphism of bicategories we get a natural equivalence of split stacks $L(F^I) \longrightarrow (LF)^I$. Since the functor "inverse image of sheaves of sets" is left exact one gets a natural isomorphism $f^{-1}((LF)^I) \xrightarrow{\sim} (f^{-1}(LF))^I$ and it remains to find, for any prestack G over $\underline{S}'$ a natural equivalence $A(G^I) \longrightarrow (AG)^I$. One has a commutative square

$$\begin{array}{ccc} G & \xrightarrow{\ a\ } & AG \\ {\scriptstyle cG}\downarrow & & \downarrow{\scriptstyle cAG} \\ G^I & \xrightarrow[a^I]{} & (AG)^I \end{array}$$

where a is the structural map of AG . According to [4] p.77 it suffices to show that a^I is "bicouvrant" [4] p.72 , which is an easy consequence of the fact that a has this property. Q.E.D. .

Corollary 1.4. Let F and F' be left exact stacks on $\underline{S}$ and $\underline{S}'$, $m: F \longrightarrow f_*(F')$ be a morphism of stacks and $m': f^*(F) \longrightarrow F'$ the morphism associated to m by the universal property of the inverse image. Then m is left exact if and only if m' is.

This is a formal consequence of (1.3).

§2.. Classifying topos of a stack.

Proposition 2.1. Let $\underline{S}$ be a left exact $\underline{U}$-site and C a prestack over $\underline{S}$ whose fibers are equivalent to categories which belong to $\underline{U}$ (C is said to be small). Let us denote by J the coarsest topology on C such that the projection $p: C \longrightarrow \underline{S}$ is a comorphism [5] III 3.1, and by $C\text{-}\underline{S}$ the category of sheaves on C for J with values in $\underline{U}$.

(1) J is defined by the pretopology whose covering families are those $(m_i : c_i \longrightarrow c), i \in I \in \underline{U}$, such that each m_i is $\underline{S}$-cartesian and such that $p(m_i), i \in I$, is a covering family.

(2) $C\text{-}\underline{S}$ is a $\underline{U}$-topos and the morphism $\pi: C\text{-}\underline{S} \longrightarrow \underline{S}$ defined by p is essential [i.e. π^* has a left adjoint $\pi_!$]. If C is left exact then $\pi_!$ is left exact.

(3) If $\underline{S}$ is a $\underline{U}$-topos and C is a stack, then the Yoneda functor $\varepsilon: C \longrightarrow C\text{-}\underline{S}$ is full and faithful and the composite $C \xrightarrow{\varepsilon} C\text{-}\underline{S} \xrightarrow{\pi_!} \underline{S}$ is equal to p.

Proof. (1) is an easy consequence of the definition of a comorphism and of the observation made in the proof of (1.1). Let $S_a, a \in A \in \underline{U}$, be a family of generators of $\underline{S}$ and $G_a, a \in A$, be a subset of $|C_{S_a}|$ which both belongs to $\underline{U}$ and contains an element of each isomorphism class of objects of the fiber C_{S_a}. The union of the G_a is a generator of the site (C,J), hence this one is a $\underline{U}$-site and $C\text{-}\underline{S}$ is a $\underline{U}$-topos. Using (1) one sees easily that for any sheaf F on $\underline{S}$, Fp is a sheaf on C hence $\pi^*(F) = Fp$, hence π^* has a left adjoint hence π is essential. The last assertion of (2) follows from the fact that when C is left exact, p is the underlying functor of a morphism of sites $\underline{S} \longrightarrow C$. The first assertion of (3) follows readily from (1) and the patching condition for morphisms in C. For any $S\varepsilon|\underline{S}|$, and any $c\varepsilon|C_S|$ one has

$$\mathrm{Hom}(\pi_!\varepsilon(c),S)=\mathrm{Hom}(\varepsilon(c),\pi^*(S))=\pi^*(S)(c)=\mathrm{Hom}(p(c),S)$$

by adjunction, Yoneda and the formula $\pi^*F = Fp$, and this concludes the proof.

2.2. Under the assumptions of (2.1), C-$\underline{S}$ is called the <u>classifying topos of the</u> (pre)<u>stack</u> C. Note that a morphism of stacks $m:C \longrightarrow C'$ is a comorphism of sites and induces a morphism of topos $m\text{-}\underline{S}:C\text{-}\underline{S} \longrightarrow C'\text{-}\underline{S}$. If m is an equivalence, then so is $m\text{-}\underline{S}$. If C is a split stack one can define a split stack C^V whose fibers are the opposites of the fibers of C. Note that the underlying category of C^V <u>is not</u> the opposite C^o of C. Let us consider the category

(1) $$B_C(\underline{S}) = St_{\underline{S}}(C^V,SH(\underline{S}))$$

of morphisms of stacks $F:C^V \longrightarrow SH(\underline{S})$, where $SH(\underline{S})$ is the split stack whose fiber at $S\varepsilon|\underline{S}|$ is the category of sheaves on $\underline{S}/S$ [equivalent to $\underline{S}/S$ since $\underline{S}$ is a topos]. One has a natural functor

(2) $$\tau^*:\underline{S} \longrightarrow B_C(\underline{S}) \quad , \quad \tau^*(S)(c) = \varepsilon(S\times p(c)) \quad ,$$

where ε is the Yoneda functor of $\underline{S}/S$.

<u>Proposition 2.3</u>. If $\underline{S}$ is a $\underline{U}$-topos and C a split stack one has an equivalence of categories

(1) $$b:B_C(\underline{S}) \longrightarrow C\text{-}\underline{S} \quad , \quad b(F)(c) = F(c)(p(c))$$

and an isomorphism of functors $b.\tau^* \xrightarrow{\sim} \pi^*$.

2.3.1. Note that this proposition proves that $B_C(\underline{S})$ is a $\underline{U}$-topos equivalent to C-$\underline{S}$ even when C is not split since one can replace C by an equivalent split stack. Furthermore, by the universal property of the associated stack, $B_C(\underline{S})$ is equivalent to $B_{C'}(\underline{S})$ when C is the stack associated to some prestack C' .

Furthermore, Lawvere and Tierney have introduced for any category-object E of the topos $\underline{S}$, the topos of objects of $\underline{S}$ provided with operations of E . One can prove that this topos is equivalent to $B_C(\underline{S})$ where C is the split prestack defined by E hence also equivalent to C'-$\underline{S}$, where C' is the stack generated by C . Thus we have three constructions of the classifying topos.

2.3.2. For any split stack D , any map $f:T \longrightarrow S$ in $\underline{S}$ and any $s \in |D_S|$ we denote by s^f the inverse image of s by f and by $s_f : s^f \longrightarrow s$ the cartesian map given by the splitting. To define b completely one must define for any $m:c \longrightarrow c'$ in C an application $b(F)(m):b(F)(c') \longrightarrow b(F)(c)$. Let $f = p(m)$, $f:S' \longrightarrow S$. Since C is split there is a canonical factorisation $c' \xrightarrow{m'} c^f \xrightarrow{c_f} c$. Since F is cartesian one has a canonical isomorphism $i:F(c^f) \to F(c)^f$ which for the values at S' (or rather $id_{S'}$) of these sheaves induces a bijection $j:F(c^f)(S') \longrightarrow F(c)(f)$ and we take for $b(F)(m)$ the composite

$$F(c)(S) \xrightarrow{f(c)(\dot{f})} F(c)(f) \xrightarrow{j^{-1}} F(c^f)(S') \xrightarrow{f(m')(S')} F(c')(S') ,$$

where $\dot{f}:f \longrightarrow id_S$ is the terminal map in $\underline{S}/S$. It is easily checked that $b(F)$ is a functor, recalling that the underlying category of C^V is not the underlying category of C^o . The sheaf axiom for $b(F)$ is verified by using (2.1(1)): for a given family $(c_i \to c)$ it is a consequence of the fact that $F(c)$ is a sheaf and

F a cartesian functor. The functoriality with respect to F is obvious. To prove that b is an equivalence one constructs explicitly a functor

(2) $\quad a:C\text{-}\underline{S} \longrightarrow B_C(\underline{S}) \quad , \quad a(G)(c)(f) = G(c^f) \quad ,$

where $c\varepsilon|F|$ and $f:T \longrightarrow p(c)$ is a map in $\underline{S}$.

Proposition 2.4. Let $f:\underline{S}' \longrightarrow \underline{S}$ be a morphism of $\underline{U}$-topos and let C be a left exact stack over $\underline{S}$. One has an equivalence of categories

(1) $\mathrm{Top}_{\underline{S}}(\underline{S}',C\text{-}\underline{S}) \longrightarrow \mathrm{Stex}_{\underline{S}}(C,f_*SH(\underline{S}'))^{o}$, where the domain is the category of morphisms of $\underline{S}$-topos $n:\underline{S}' \longrightarrow C\text{-}\underline{S}$, where $f_*SH(\underline{S}')$ is the direct image by f of the stack of sheaves over $\underline{S}'$ [its fiber at $S\varepsilon|\underline{S}|$ is the category of sheaves over $\underline{S}'/f^*(S)$] and where the codomain is the opposite of the category of left exact morphisms of stacks $C \longrightarrow f_*SH(\underline{S}')$.

Since C is left exact and $\varepsilon:C \longrightarrow C\text{-}\underline{S}$ full and faithful, a morphism of topos $n:\underline{S}' \longrightarrow C\text{-}\underline{S}$ is nothing but a left exact functor $n^{-1}:C \longrightarrow \underline{S}'$, $n^{-1}=n^*\varepsilon$. Furthermore, since C is left exact there exists a cartesian section p^{-1} of C whose value at $S\varepsilon|\underline{S}|$ is the terminal object of the fiber C_S and p^{-1} is a morphism of sites defining $\pi:C\text{-}\underline{S} \longrightarrow \underline{S}$ since $\pi^*F = Fp$ for any sheaf F on $\underline{S}$. Hence an isomorphism of morphisms of topos $i:\pi n \xrightarrow{\sim} f$ is nothing but an isomorphism $i^{-1}:n^{-1}p^{-1} \xrightarrow{\sim} f^*$. In other words the category $\mathrm{Top}_{\underline{S}}(\underline{S}',C\text{-}\underline{S})^{o}$ is equivalent to the category M of pairs $(n^{-1}:C \longrightarrow \underline{S}', i^{-1}:n^{-1}p^{-1} \xrightarrow{\sim} f)$ where n^{-1} is continuous and left exact. Let $Ar(\underline{S}')$ be the category whose objects are arrows of $\underline{S}'$ and let $b:Ar(\underline{S}') \longrightarrow \underline{S}'$, $b(X \longrightarrow Y) = Y$. Since every object $c\varepsilon|C|$ determines a terminal map $c \longrightarrow p^{-1}(p(c))$, a pair (n^{-1},i^{-1}) can be viewed as a functor $n':C \longrightarrow Ar(\underline{S}')$ such that

$bn' = f^*p$ and which is left exact [the continuity condition disappears by (2.1 (1))]. Since b makes a stack over $\underline{S}'$ out of the category $Ar(\underline{S}')$, by the very definition of the direct image of a stack, n' is nothing but a functor $n'': C \longrightarrow f_*Ar(\underline{S}')$ and, since n' is left exact, n'' is $\underline{S}$-cartesian and left exact, hence an object of $Stex_{\underline{S}}(C, Ar(\underline{S}'))$. The conclusion follows since $Ar(\underline{S}')$ is equivalent to $SH(\underline{S}')$.

According to the proof, the morphism of topos $n: \underline{S}' \longrightarrow C\text{-}\underline{S}$ which corresponds to a left exact morphism of stacks $n'': C \longrightarrow f_*Ar(\underline{S}')$ is characterized up to unique isomorphism by the equality $n^*\varepsilon = dqn''$

(2) $$C \xrightarrow{n''} f_*Ar(\underline{S}') \xrightarrow{q} Ar(\underline{S}') \xrightarrow{d} \underline{S}' ,$$

where q is the first projection of $f_*Ar(\underline{S}') = Ar(\underline{S}') \times_{\underline{S}'} \underline{S}$, d the "domain functor" and ε the Yoneda functor.

Corollary 2.5. If C is left exact one has an equivalence [(1)]

(1) $$Top_{\underline{S}}(\underline{S}', C\text{-}\underline{S}) \longrightarrow Stex_{\underline{S}'}(f^*(C), SH(\underline{S}'))^{\circ} .$$

This follows immediately from (2.4), (1.4) and the universal property of the inverse image $f^*(C)$ of C . This gives the universal property of $C\text{-}\underline{S}$ in the bicategory of $\underline{S}$-topos.

Corollary 2.6. Let $C' = f^*(C)$. One has a commutative square of morphisms of topos

(1) $$\begin{array}{ccc} C\text{-}\underline{S} & \xleftarrow{C\text{-}f} & C'\text{-}\underline{S}' \\ \downarrow & & \downarrow \\ \underline{S} & \xleftarrow[f]{} & \underline{S}' \end{array}$$

which is bicartesian.

---

(1) $Stex_{\underline{S}}(\ , \ )$ stands for "category of left exact morphisms of stacks".

This means that for any morphism of topos $g:\underline{S}'' \longrightarrow \underline{S}'$ the functor given by composition with C-f

(2) $$\mathrm{Top}_{\underline{S}'}(\underline{S}'', C'\text{-}\underline{S}') \longrightarrow \mathrm{Top}_{\underline{S}}(\underline{S}'', C\text{-}\underline{S})$$

is an equivalence. By the very definition of C' [4] p.87, one has a commutative square

(3) $$\begin{array}{ccc} C & \xrightarrow{\phi^{-1}} & C' \\ {\scriptstyle p}\downarrow & & \downarrow{\scriptstyle p'} \\ \underline{S} & \xrightarrow{f^*} & \underline{S}' \end{array}$$

where ϕ^{-1} is cartesian. Furthermore ϕ^{-1} is left exact by (1.3). By (1.4) and the univsal property of $C' = f^*(C)$, for any $g:\underline{S}'' \longrightarrow \underline{S}'$, the functor

(4) $\mathrm{Stex}_{\underline{S}'}(C', g_*SH(\underline{S}'')) \longrightarrow \mathrm{Stex}_{\underline{S}}(C, f_*g_*SH(\underline{S}''))$, $u \longmapsto u\phi^{-1}$, is an equivalence. By (2.4) the proof is now an exercise about universal properties in bicategories.

§3. Generating stack of a $\underline{U}$-topos.

The question of defining a relative notion of generators has been raised by Lawvere and Tierney. We propose here an answer in the language of $\underline{U}$-topos. It is clear that Prop.(3.3) is still valid when working in their framework and that (3.2) is not.

Definition 3.1. Let $f:\underline{X} \longrightarrow \underline{S}$ be a morphism of $\underline{U}$-topos. A generating stack of f is a full substack C of $F = f_*(\mathrm{Ar}(\underline{X}))$ which is small (2.1) and such that, for any $S \in |\underline{S}|$ and any $x \in |F_S|$, there exists a covering family $(S_i \longrightarrow S)$, $i \in I$, in $\underline{S}$ and for each $i \in I$ a covering family $(c_{i,j} \longrightarrow x_i)$ in the fiber $F_S = \underline{X}/f^*(S)$, with $c_{i,j} \in |C|$, where x_i is the inverse image of x by $S_i \longrightarrow S$. A generating stack C is said to be left exact if C and the inclusion functor $i:C \longrightarrow F$ are left exact.

Let us recall that the category of arrows of $\underline{X}$ provided with the codomain functor $Ar(\underline{X}) \longrightarrow \underline{X}$ is a stack. Hence its direct image F is a stack whose fiber at $S \varepsilon |\underline{S}|$ is the topos $\underline{X}/f^*(S)$ and the inverse image functor $F_u : F_S \longrightarrow F_{S'}$ associated to a map $u:S' \longrightarrow S$ in $\underline{S}$ is nothing but pull-back along $f^*(u):f^*(S') \longrightarrow f^*(S)$. Hence F is a left exact stack and the condition that a full substack C of F is left exact is that each fiber C_S is stable by finite limits in the fiber F_S .

<u>Proposition 3.2</u>. Any $\underline{S}$-topos admits a left exact generating stack.

Let us choose a generator $(S_i), i \varepsilon I \varepsilon \underline{U}$, of $\underline{S}$ and for each $i \varepsilon I$ a full subcategory C_i of F_{S_i} stable by finite limits, generating F_{S_i} and equivalent to a category which belongs to $\underline{U}$. Let us define C as the full subcategory of F whose objects of projection $S \varepsilon |\underline{S}|$ are those $x \varepsilon |F_S|$ such that there exists a covering family $(c_a : S_a \longrightarrow S)$, such that each S_a is one of the S_i and the inverse image of x by c_a is isomorphic to an object of C_i . This condition being local on $\underline{S}$, it is clear that C is a full substack of F and even a left exact one since F is left exact. Furthermore C is small since for each $S \varepsilon |\underline{S}|$ the set of classes of equivalent covering families $(S_a \longrightarrow S)$ as above belongs to $\underline{U}$. Eventually C is a generating stack since any $S \varepsilon |\underline{S}|$ can be covered by the S_i .

<u>Proposition 3.3</u>. Let $\underline{S}$ be a $\underline{U}$-topos and C a generating stack of an $\underline{S}$-topos $f:\underline{X} \longrightarrow \underline{S}$. Then C-$\underline{S}$ is an $\underline{S}$-topos and there exists an $\underline{S}$-morphism of topos $n:\underline{X} \longrightarrow C\text{-}\underline{S}$ such that $n_*:\underline{X} \longrightarrow C\text{-}\underline{S}$ is full and faithful [in other words $\underline{X}$ is a subtopos of C-$\underline{S}$] .

3.3.1. We note first that since C is small, C-$\underline{S}$ is a $\underline{U}$-topos. Furthermore there exists a left exact generating stack C' of $\underline{X}$ containing C and such that each object of C' is a finite limit of objects of C. Hence the inclusion $C \longrightarrow C'$ induces an equivalence between the $\underline{S}$-topos C-$\underline{S}$ and C'-$\underline{S}$ and this fact allows us to assume that C <u>is left exact</u>. Since the inclusion $i:C \longrightarrow F$, $F = f_*Ar(\underline{X})$, is left exact one has an $\underline{S}$-morphism $n:\underline{X} \longrightarrow C$-$\underline{S}$, (2.4), whose inverse image functor $n^*:C$-$\underline{S} \longrightarrow \underline{X}$ is such that its composition with the Yoneda functor $\varepsilon:C \longrightarrow C$-$\underline{S}$ is equal to the composite of

$$(1) \qquad C \xrightarrow{i} F \xrightarrow{q} Ar(\underline{X}) \xrightarrow{d} X \quad , \quad (2.4(2)) .$$

For any $c \in |C|$ and any $X \in |\underline{X}|$ one has $n_*(X)(c) = Hom(\varepsilon(c), n_*(X)) = Hom(n^*\varepsilon(c), X) = Hom(dqi(c), X) = Hom_S(i(c), X \times f^*(S))$ where the last set of morphisms is taken in the fiber $\underline{X}/f^*(S)$ of F with $S = p(c)$, and the last equality sign is justified by the definition of F as a fibered product. Hence the formula

$$(2) \qquad n_*:\underline{X} \longrightarrow C\text{-}\underline{S} \quad , \quad n_*(X)(c) = Hom_S(i(c), X \times f^*(S)) \quad , \quad S = p(c) .$$

3.3.2. To prove that n_* is full and faithful we will first compose it with the inverse $a:C$-$\underline{S} \longrightarrow B_C(\underline{S})$ of (2.3(1))

$$(3) \quad an_*:\underline{X} \longrightarrow B_C(\underline{S}) \quad , \quad an_*(X)(c) = \underline{Hom}_S(i(c), X \times f^*(S)) \quad , \quad S = p(c) \quad , \quad c \in |C| \quad ,$$

the above formula being justified by (2.3(2)), where $\underline{Hom}_S(u,v)$ stands for the sheaf (over S) of S-morphisms between the objects u and v of the fiber at S of the stack F. Let us prove that (3) is the effect on the fibers at the terminal object of $\underline{S}$ of a morphism of stacks

(4) $m: F \longrightarrow ST(C^V, SH(\underline{S}))$,

where $ST(A,B)$ stands for the (split) stack of morphisms of stacks between A and B [internal Hom in the bicategory of stacks [4] p.57, 77] , whose fiber at $S \in |\underline{S}|$ is the category of morphisms $A/S \longrightarrow B/S$ of stacks over $\underline{S}/S$. We obtain (4) by composition of

(5) $F \xrightarrow{\ y\ } ST(F^V, SH(\underline{S})) \xrightarrow{\ j\ } ST(C^V, SH(\underline{S}))$

where j is induced by composition with $i: C \longrightarrow F$ and where y is a "relative Yoneda functor" defined by

(6) $y(a)(b) = \underline{Hom}_S(b, a^f)$,

where $f: T \longrightarrow S$ is a map in $\underline{S}$ and $a \in |F_S|$, $b \in |F_T|$. One should note that the restriction of y to the terminal fiber of F is also the restriction of the composite $F \xrightarrow{\varepsilon} F\text{-}\underline{S} \xrightarrow{a} B_F(\underline{S})$, (2.1(3)), (2.3(2)). By localisation it follows that the restriciton of y to each fiber is full and faithful hence y is such. On the other hand, since any object of F can be covered for the canonical topology of F by objects of $i(C)$ and since i is full and faithful it is easy to show that j is also full and faithful and the conclusion follows.

Proposition 3.4. Fibered products exist in the bicategory of $\underline{U}$-topos.

According to (3.2) and (3.3) any morphism of $\underline{U}$-topos $\underline{X} \longrightarrow \underline{S}$ can be factored in $\underline{X} \xrightarrow{n} C\text{-}\underline{S} \xrightarrow{\pi} \underline{S}$ where n_* is full and faithful and where C is a left exact small stack over $\underline{S}$. By (2.6) the pullback of π along any morphism of $\underline{U}$-topos $f: \underline{S}' \longrightarrow \underline{S}$ exists. On the other hand the pull-back of n along any morphism of $\underline{U}$-topos $y: \underline{Y} \longrightarrow C\text{-}\underline{S}$ exists because $\underline{X}$ is a subtopos of $C\text{-}\underline{S}$ hence is defined by some topology J on $C\text{-}\underline{S}$ and it suffices to take as a pullback the subtopos of $\underline{Y}$ defined by the finest topology

J' on $\underline{Y}$ such that the inverse image functor $y^*:C\text{-}\underline{S} \longrightarrow \underline{Y}$ is continuous. The conclusion follows by transitivity of pullback in a bicategory.

Bibliography

[1] P. DELIGNE and D. MUMFORD, The irreducibility of the space of curves of given genus, Publications mathématiques de l'I.H.E.S., No. 36,1969, p. 75-110.

[2] J. GIRAUD, Méthode de la descente, Mémoire de la Société Mathématique de France, 1964.

[3] __________, Effacement d'une classe de cohomologie de degré 2, C.R. Ac. Sc. 265, 229-231, (1967).

[4] __________, Cohomologie non abélienne, Springer, 1971.

[5] A. GROTHENDIECK, Séminaire de Géométrie Algébrique de l'I.H.E.S., Cohomologie étale des schémas, 1964.

[6] F.W. LAWVERE, Quantifiers and sheaves, Congrès International des Mathématiciens, Nice 1970.

[7] D. MUMFORD, Picard group of moduli problems, Proc. Conf. on Arith. Alg. Geom. at Purdue, 1963.

DEDUCTIVE SYSTEMS AND CATEGORIES
III. CARTESIAN CLOSED CATEGORIES, INTUITIONIST PROPOSITIONAL CALCULUS, AND COMBINATORY LOGIC

by

Joachim Lambek

INTRODUCTION

In this paper we discuss two connections between cartesian closed categories and logic. On the one hand, the positive intuitionist calculus is used to construct free cartesian closed categories. On the other hand, cartesian closed categories bear a close relationship to combinatory logic with types. Thus category theory acts as a bridge between two branches of logic and makes explicit the formal analogy between theorems in intuitionist logic and the types of functionality, which has already been pointed out by Curry and Feys. Moreover, Schönfinkel's original program to base all of mathematics on combinatory logic can now be said to have been carried out by Lawvere in his many articles pursuing a categorical foundation of mathematics.

Most of the present material was first presented at the Battelle Institute Conference on Category Theory in 1968, but was not included in the proceedings of that conference. While functional completeness was then expressed in terms of "multicategories", the present exposition favours "ontologies" which are here introduced for the first time.

This paper is the third in a series, but familiarity with "Deductive systems and categories, I and II" is not presumed.

1. CARTESIAN CLOSED CATEGORIES

A *cartesian closed category* $(\mathfrak{A}, T, \wedge, \Leftarrow, \alpha, \beta, \gamma)$ consists of a category $\mathfrak{A}$, an object T of $\mathfrak{A}$, bifunctors $\wedge : \mathfrak{A} \times \mathfrak{A} \longrightarrow \mathfrak{A}$ and $\Leftarrow : \mathfrak{A} \times \mathfrak{A}^{op} \longrightarrow \mathfrak{A}$, and three natural isomorphisms

$$\alpha(A) : [A, T] \simeq T,$$

$$\beta(C, A, B) : [C, A \wedge B] \simeq [C, A] \times [C, B],$$

$$\gamma(C, A, B) : [C, A \Leftarrow B] \simeq [C \wedge B, A].$$

These isomorphisms assert that

(α) T is a terminal object,

(β) $\wedge$ is the right adjoint of the diagonal functor $\Delta : \mathfrak{A} \longrightarrow \mathfrak{A} \times \mathfrak{A}$,

(γ) $(\ ) \Leftarrow B$ is the right adjoint of the functor $(\ ) \wedge B : \mathfrak{A} \longrightarrow \mathfrak{A}$, for each object B of $\mathfrak{A}$.

Familiar examples are the categories *Ens* of small sets and *Cat* of small categories. In fact these examples suggest the more usual notation

$$T = \underline{1}, \quad A \wedge B = A \times B, \quad A \Leftarrow B = A^B,$$

which we eschew for typographical reasons, and also because of the comparison with logic that is always at the back of our minds. Another kind of example is obtained if we assume that between any two objects of $\mathfrak{A}$ there is at most one map. $\mathfrak{A}$ is then essentially a pre-ordered set (we do not assume antisymmetry), and a cartesian closed pre-ordered set is what might also be called a "relatively pseudo complemented pre-semilattice", but we shall refrain from spelling out the separate meanings of these words. It is now fashionable to call this a 'Heyting algebra'.

Everybody knows that the existence of a right adjoint U to a given functor $F : \mathfrak{A} \longrightarrow \mathfrak{B}$ is equivalent to the universal mapping property: for each object B of $\mathfrak{B}$ there is an object $U(B)$ of $\mathfrak{A}$ and a map $\epsilon(B) : FU(B) \longrightarrow B$ so that for each object A of $\mathfrak{A}$ and each map b $F(A) \longrightarrow B$ there exists a unique map $b^* : A \longrightarrow U(B)$ such that $\epsilon(B)F(b^*) = b$.

The uniqueness of b^* asserts that, for any $a : A \longrightarrow U(B)$,

$$\epsilon(B)a = b \Rightarrow a = b^*.$$

This statement may be expressed as an equation: $(\epsilon(B)a)^* = a$, and this fact will be important to us.

We also point out that, for each map $b : B' \longrightarrow B$ in $\mathfrak{B}$, $U(b) : U(B') \longrightarrow U(B)$ is the unique map which renders the following diagram commutative:

$$\begin{array}{ccc} FU(B) & \xleftarrow{\quad FU(b) \quad} & FU(B') \\ \Big\downarrow {\scriptstyle \epsilon(B)} & & \Big\downarrow {\scriptstyle \epsilon(B')} \\ B & \xleftarrow[\quad b \quad]{} & B' \end{array}$$

In other words, $U(b) = (b\,\epsilon(B'))^*$.

We shall apply these ideas first to the functor Δ and then to the functor $(\ )\wedge B$ mentioned above. Replacing $F : \mathfrak{A} \longrightarrow \mathfrak{B}$ by $\Delta : \mathfrak{A} \longrightarrow \mathfrak{A} \times \mathfrak{A}$, where $\Delta(A) = (A, A)$, we see that $\epsilon(A_1, A_2) = (p(A_1, A_2), q(A_1, A_2))$, where $p(A_1, A_2) : A_1 \wedge A_2 \longrightarrow A_1$ and $q(A_1, A_2) : A_1 \wedge A_2 \longrightarrow A_2$ are the so-called *projection* maps. Replacing F by $(\ )\wedge B : \mathfrak{A} \longrightarrow \mathfrak{A}$, for a fixed object B of $\mathfrak{A}$, we see that $\epsilon(A) = e(A, B) : (A \Leftarrow B) \wedge B \longrightarrow A$, the so-called *evaluation* map.

We then obtain the following alternative presentation of a cartesian closed category as a 10-tuple $(\mathfrak{A}, T, \wedge, \Leftarrow, 0, p, q, \langle\rangle, e, {}^*)$, where now $\wedge$ and $\Leftarrow$ are only given as object functions, $0, p, q$ and e are families of maps, and $\langle\rangle$ and * are devices for creating new maps from old, as follows:

$$0(A) : A \longrightarrow T \qquad p(A, B) : A \wedge B \longrightarrow A \qquad e(A, B) : (A \Leftarrow B) \wedge B \longrightarrow A$$

$$q(A, B) : A \wedge B \longrightarrow B$$

$$\frac{a : C \longrightarrow A \qquad b : C \longrightarrow B}{\langle a, b \rangle : C \longrightarrow A \wedge B} \qquad \frac{h : C \wedge B \longrightarrow A}{h^* : C \longrightarrow A \Leftarrow B}$$

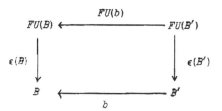

Moreover, these symbols are subject to the following equations:

$$f = 0, \quad p\langle a, b\rangle = a, \qquad e\langle h^{*}p, q\rangle = h,$$

$$q\langle a, b\rangle = b, \qquad (e\langle fp, q\rangle)^{*} = f.$$

$$\langle pf, qf\rangle = f,$$

In writing these equations, we have abbreviated $0(A)$ by 0, $p(A,B)$ by p, etc. Furthermore, it is tacitly understood that the two sides of an equation have the same domain and codomain. For example, the equation $f = 0$ really asserts that, for any map $f : A \longrightarrow T$, $f = 0(A)$.

The second presentation of cartesian closed categories is of course more cumbersome and less transparent than the first. However, it serves a specific purpose, to show the equational nature of this concept. In fact, it can be shown that the category $\mathcal{Cart}$ of small cartesian closed categories is equational in a technical sense over $\mathcal{Cat}$. We shall not discuss this here, but we shall construct a left adjoint to the obvious forgetful functor $\mathcal{Cart} \longrightarrow \mathcal{Cat}$.

The functoriality of $\wedge$ and $\Leftarrow$ is expressed by the definitions

$$f \wedge g = \langle fp, gq\rangle, \qquad f \Leftarrow g' = (fe\langle p, g'q\rangle)^{*},$$

where $f : A' \longrightarrow A$, $g : B' \longrightarrow B$, and $g' : B \longrightarrow B'$. In fact, the definition of $f \wedge g$ has already been used in stating the equations pertaining to e and *.

2. POSITIVE INTUITIONIST PROPOSITIONAL LOGIC AS A DEDUCTIVE SYSTEM

Let $\mathfrak{X}$ be a given small category. We shall discuss a deductive system $D(\mathfrak{X})$ depending on $\mathfrak{X}$. $D(\mathfrak{X})$ will be a language made up of the following list of primitive symbols:

$$\rightarrow \; (\;) \; , \; : \; T \; \wedge \; \Leftarrow \; 0 \; p \; q \; \langle\rangle \; e \; {}^{*}$$

as well as names of objects and maps in $\mathfrak{X}$.

In talking about expressions in this language, we shall omit quotation marks (or rather Quine's corners). Thus, if A and B are expressions in the language, $(A \wedge B)$ is the expression consisting of "(", followed by A, followed by "$\wedge$", followed by B, followed by ")". In particular, we do not distinguish between the objects and maps of $\mathfrak{X}$ on the one side and their names on the other.

The *terms* of $D(\mathfrak{X})$ are defined inductively:

(1) Every object of $\mathfrak{X}$ is a term.

(2) T is a term.

(3) If A and B are terms then so are $(A \wedge B)$ and $(A \Leftarrow B)$.

(4) Nothing else is a term.

By a *formula* in $D(\mathfrak{X})$ we mean an expression of the form $A \longrightarrow B$, where A and B are terms. Some authors would call our terms "formulas" and our formulas "sequents".

By a *proof* of a formula $A \longrightarrow B$ we mean an expression of the form $P: A \longrightarrow B$ defined inductively as follows.

(1) The following expressions are proofs:

$1(A): A \longrightarrow A$,

$0(A): A \longrightarrow T$,

$p(A, B): (A \wedge B) \longrightarrow A$,

$q(A, B): (A \wedge B) \longrightarrow B$,

$e(A, B): ((A \Leftarrow B) \wedge B) \longrightarrow A$.

(2) Every map $f: X \longrightarrow Y$ of $\mathfrak{X}$ is a proof.

(3) If $P: A \longrightarrow B$ and $Q: B \longrightarrow C$ are proofs, so is $(QP): A \longrightarrow C$.

(4) If $P: C \longrightarrow A$ and $Q: C \longrightarrow B$ are proofs, so is $\langle P, Q \rangle : C \longrightarrow A \wedge B$.

(5) If $P: (C \wedge B) \longrightarrow A$ is a proof, so is $P^{*}: C \longrightarrow (A \Leftarrow B)$.

(6) Nothing else is a proof.

We say that $A \to B$ is a *theorem* if there exists a proof $P: A \to B$.

We may think of (1) and (2) as stating the axioms of our deductive system, while (3), (4) and (5) are the rules of inference. Each proof may be illustrated by a tree. For example,

$$\langle q(A,B),\ p(A,B)\rangle : (A \wedge B) \to (B \wedge A)$$

is illustrated by

$$\frac{A \wedge B \to B \qquad\qquad A \wedge B \to A}{A \wedge B \to B \wedge A}\ .$$

However, some information is lost when replacing a proof by its tree. Somewhat pedantically we wish to distinguish between the axioms $1(X): X \to X$ and $1_X: X \to X$, where 1_X is an identity map in $\mathfrak{X}$, also between $p(T,B): T \wedge B \to T$ and $0(T): T \wedge B \to T$, although these proofs will later turn out to be equivalent in a certain sense. More important are the distinctions between proofs $p(A,A)$: $A \wedge A \to A$ and $q(A,A): A \wedge A \to A$, and between proofs $f: X \to Y$ and $g: X \to Y$ when f and g are distinct maps in $\mathfrak{X}$.

Suppose $\mathfrak{X}$ is a discrete category with objects $p, q, \ldots$, that is, $\mathfrak{X}$ is essentially the set $\{p, q, \ldots\}$. Then $D(\mathfrak{X})$ is the deductive system which describes the *positive intuitionist propositional calculus*. It is positive, because only the connectives $\wedge$ and $\Leftarrow$ occur, in addition to T. It is intuitionist, because, e.g., the formula $(p \Leftarrow (q \Leftarrow p)) \to p$ is not provable in $D(\mathfrak{X})$, although it is a tautological implication.

As a second example, suppose $\mathfrak{X}$ is a pre-ordered set. Then $D(\mathfrak{X})$ may be used to construct the free cartesian closed pre-ordered set $F(\mathfrak{X})$ generated by $\mathfrak{X}$: its elements are the terms of $D(\mathfrak{X})$, and $A \leq B$ in $F(\mathfrak{X})$ provided $A \to B$ is a theorem in $D(\mathfrak{X})$.

3. FREE CARTESIAN CLOSED CATEGORIES

Our present purpose is to construct the free cartesian closed category $F(\mathfrak{X})$ generated by a given small category $\mathfrak{X}$. To make this precise, let *Cart* be the category whose objects are small cartesian closed categories and whose maps, call them *strict maps*, are functors which preserve the structure strictly. Thus we require of such a map G that it be a functor and that $G(T) = T$, $G(A \wedge B) = G(A) \wedge G(B), \ldots, G(p(A,B)) = p(G(A), G(B)), \ldots, G(\langle f, g \rangle) = \langle G(f), G(g) \rangle, \ldots$ Functors which preserve the structure strictly, not just up to isomorphism, may not occur much in nature, but they are useful for what we have in mind. One may drop the insistance on strict preservation, but then one must use "free" in an unorthodox fashion.

We claim that the obvious forgetful functor $U : Cart \longrightarrow Cat$ has a left adjoint F. We shall construct a cartesian closed category $F(\mathfrak{X})$ together with a functor $\eta(\mathfrak{X}) : \mathfrak{X} \longrightarrow UF(\mathfrak{X})$ such that, for every cartesian closed category $(\mathfrak{C}, T, \wedge, \Leftarrow, \ldots)$ and every functor $G : \mathfrak{X} \longrightarrow \mathfrak{C}$, there exists a unique strict map G' in *Cart* such that $U(G')\,\eta(\mathfrak{X}) = G$.

CONSTRUCTION 1. The objects of $UF(\mathfrak{X})$ are the terms of $D(\mathfrak{X})$. The maps of $UF(\mathfrak{X})$ are equivalence classes of proofs in $D(\mathfrak{X})$ according to a certain equivalence relation $\equiv$ to be described below: the equivalence class $[P]$ of the proof $P : A \longrightarrow B$ is regarded as a map $[P] : A \longrightarrow B$. In particular, $[1(A)] : A \longrightarrow A$ is the identity map, and the composition of $[P] : A \longrightarrow B$ and $[Q] : B \longrightarrow C$ is defined by $[Q][P] = [QP]$. The functor $\eta(\mathfrak{X}) : \mathfrak{X} \longrightarrow UF(\mathfrak{X})$ is defined thus:

$$\eta(\mathfrak{X})(X) = X, \quad \eta(\mathfrak{X})(f) = [f].$$

Moreover, $F(\mathfrak{X})$ is given the cartesian closed structure by utilizing T, $\wedge$, and $\Leftarrow$, and putting

$$0(A) = [0(A)], \quad p(A,B) = [p(A,B)], \quad q(A,B) = [q(A,B)],$$

$$\langle [P],[Q]\rangle = [\langle P,Q\rangle], \quad [P]^* = [P^*].$$

Finally, we define the equivalence relation $\equiv$ between proofs $P : A \longrightarrow B$ and $Q : A \longrightarrow B$ as the smallest relation satisfying the following three sets of conditions:

(I) Like any decent equivalence relation, $\equiv$ should satisfy the reflexive, symmetric, transitive, and substitution laws:

$P \equiv P$;

if $P \equiv Q$ then $Q \equiv P$;

if $P \equiv Q$ and $Q \equiv R$ then $P \equiv R$;

if $P \equiv Q$ then $P^* \equiv Q^*$;

if $P \equiv Q$ and $R \equiv S$ then $\langle P,R\rangle \equiv \langle Q,S\rangle$.

(II) To ensure that $UF(\mathfrak{X})$ is a category and that $\eta(\mathfrak{X})$ is a functor, we require that

$P\,1(A) \equiv P$ for each proof $P : A \longrightarrow B$;

$1(B)P \equiv P$ for each proof $P : A \longrightarrow B$;

$R(QP) \equiv (RQ)P$ for all proofs $P : A \longrightarrow B$, $Q : B \longrightarrow C$, $R : C \longrightarrow D$;

$1_X \equiv 1(X)$ for each object X of $\mathfrak{X}$;

$gf \equiv g \circ f$ for all maps $f : X \longrightarrow Y$ and $g : Y \longrightarrow Z$ in $\mathfrak{X}$, where $\circ$ denotes composition in $\mathfrak{X}$.

(III) $F(\mathfrak{X})$ should satisfy the equations of a cartesian closed category, thus

$P \equiv 0(A)$ for each proof $P : A \longrightarrow T$;

$p(A,B)\langle P,Q\rangle \equiv P$;

$q(A,B)\langle P,Q\rangle \equiv Q$;

$\langle p(A,B)R,\ q(A,B)R\rangle \equiv R$;

$e(A,B)\langle R^*p(C,B),\ q(C,B)\rangle \equiv R$;

$(e(A,B)\langle Pp(C,B),\ q(C,B)\rangle)^* \equiv P$.

The proof that the above construction has the desired property is similar to corresponding proofs in "Deductive systems and categories, I and II", and will therefore be omitted.

The above equivalence relation between proofs could also be described by means of a new deductive system $D'(\mathfrak{X})$. The terms of $D'(\mathfrak{X})$ are the proofs of $D(\mathfrak{X})$, and the formulas of $D'(\mathfrak{X})$ have the form $P \equiv Q$, where $P: A \longrightarrow B$ and $Q: A \longrightarrow B$ are proofs in $D(\mathfrak{X})$. Axioms and rules of inference for $D'(\mathfrak{X})$ correspond to three sets of conditions I, II and III above. For example, the first two conditions give rise to the axiom $P \equiv P$ and to the rule of inference $\frac{P \equiv Q}{Q \equiv P}$. One would then say that P and Q are equivalent if $P \equiv Q$ is provable in $D'(\mathfrak{X})$.

It is of course tedious to describe the equivalence relation between proofs by the large sets of conditions I, II and III. I have asserted elsewhere for similar deductive systems that two proofs of $A \longrightarrow B$ are equivalent if and only if they have the same "generality", and this idea has been adapted to the present set up by Manfred Szabo. For example, $p(A, A): A \wedge A \longrightarrow A$ and $q(A, A): A \wedge A \longrightarrow A$ have different generality, because they generalize to $p(X, Y): X \wedge Y \longrightarrow X$ and $q(X, Y): X \wedge Y \longrightarrow Y$ respectively.

While it is intuitively clear what is meant by "generality", my earlier attempts to make this concept precise were faulty, as was pointed out by Manfred Szabo and, more recently, by Max Kelly.

4. COMBINATORY LOGIC

The propositional calculus is usually presented without the deduction symbol $\rightarrow$. We may ask what happens to a cartesian closed category if we remove all maps, except maps of the form $T \rightarrow C$. It should be possible to recapture the category, in view of the natural isomorphism $[A,B] \cong [T, B \Leftarrow A]$, which we shall now make explicit.

To each map $f: A \rightarrow B$ there corresponds a map $f^{)}: T \rightarrow B \Leftarrow A$ defined by $f^{)} = (fq(T,A))^{*}$, which construction is illustrated by the following tree:

$$\dfrac{\dfrac{T \wedge A \rightarrow A \qquad A \xrightarrow{f} B}{T \wedge A \rightarrow B}}{T \rightarrow B \Leftarrow A}$$

On the other hand, to each map $g: T \rightarrow B \Leftarrow A$ there corresponds a map $g^{(}$ defined by $g^{(} = e(B,A)\langle g\,0(A), 1(A)\rangle$, which is illustrated by the tree

$$\dfrac{\dfrac{\dfrac{A \rightarrow T \qquad T \xrightarrow{g} B \Leftarrow A}{A \rightarrow B \Leftarrow A} \qquad A \rightarrow A}{A \rightarrow (B \Leftarrow A) \wedge A} \qquad (B \Leftarrow A) \wedge A \rightarrow B}{A \rightarrow B}$$

It is easily seen that $f^{)(} = f$ and $g^{()} = g$.

We shall define an *ontology* as consisting of the following data:

(1) a class $\mathcal{T}$ of *types*;

(2) a class $\mathcal{E}$ of *entities*;

(3) a binary relation $\in$ between entities and types (we read "$a \in A$" to say that a has type A);

(4) a given type T;

(5) two binary operations $\wedge$ and $\Leftarrow$ between types;

(6) a binary operation $^{\langle}$ on entities, called *application*, such that $f^{\langle} b \in A$ if $f \in A \Leftarrow B$ and $b \in B$;

(7) seven families of entities as follows:

$$I_A \in A \Leftarrow A,$$

$$K_{A,B} \in (A \Leftarrow B) \Leftarrow A,$$

$$S_{A,B,C} \in ((A \Leftarrow C) \Leftarrow (B \Leftarrow C)) \Leftarrow ((A \Leftarrow B) \Leftarrow C),$$

$$P_{A,B} \in A \Leftarrow (A \wedge B),$$

$$Q_{A,B} \in B \Leftarrow (A \wedge B),$$

$$R_{A,B} \in ((A \wedge B) \Leftarrow B) \Leftarrow A,$$

$$0 \in T.$$

(8) Moreover, we require the following set of equations:

(i) $I^{\langle} a = a$, where $a \in A$;

(ii) $(K^{\langle} a)^{\langle} b = a$, where $a \in A$, $b \in B$;

(iii) $((S^{\langle} f)^{\langle} g)^{\langle} c = (f^{\langle} c)^{\langle} (g^{\langle} c)$, where $f \in (A \Leftarrow B) \Leftarrow C$, $g \in (B \Leftarrow C)$, $c \in C$;

(iv) $P^{\langle} ((R^{\langle} a)^{\langle} b) = a$, where $a \in A$, $b \in B$;

(v) $Q^{\langle} ((R^{\langle} a)^{\langle} b) = b$, where $a \in A$, $b \in B$;

(vi) $(R^{\langle} (P^{\langle} c))^{\langle} (Q^{\langle} c) = c$, where $c \in A \wedge B$;

(vii) $0 = t$, where $t \in T$;

(viii) $(S^{\langle} (K^{\langle} c))^{\langle} (K^{\langle} d) = K^{\langle} (c^{\langle} d)$;

(ix) $(S^{\langle} (K^{\langle} c))^{\langle} I = c$.

Note that (i), for example, is a family of equations, one for each pair (a, A) such that $a \in A$.

(9) Finally we require a finite set of families of equations:

$$u_i = v_i \qquad (i = 1, 2, \ldots, n),$$

which will be specified later. The u_i and v_i are made up from I, K, S, P, Q, R and 0 by means of application.

Why such a monstrous definition? The entities I, K and S, and the equations (i), (ii) and (iii) were already given by Schönfinkel, and appear in every book on Combinatory Logic. (See Schönfinkel, Church, Rosenbloom, Curry and Feys.) The remaining entities and equations (iv) to (vii) become necessary if we wish to account for the types T and $A \wedge B$. The equations (9) go back to Rosser, their rôle will be made clear later. Our aim is to show that the concept of an ontology is equivalent to that of a cartesian closed category.

Forgetting the additional structure, we may associate with the ontology $\mathfrak{O}$ a binary relation $(\mathcal{E}, \in, \mathfrak{T})$, call it $U(\mathfrak{O})$. Conversely, given any binary relation $\mathfrak{R} = (X, R, Y)$ with $R \subseteq X \times Y$, we may associate with it the "free" ontology $F(\mathfrak{R})$ generated by $\mathfrak{R}$. This may be done in a way similar to the construction of the free cartesian closed category generated by a given small category. Only this time one would formulate the positive propositional calculus in the usual way, without the deduction symbol. We shall not dwell on this construction, as we are more interested in another universal construction.

5. POLYNOMIAL ONTOLOGIES

Given an ontology $\mathfrak{O}$, we wish to adjoin an indeterminate x of type A, that is, we wish to construct the *polynomial ontology* $\mathfrak{O}[x]$. This should come equipped with a map $\eta_x : \mathfrak{O} \longrightarrow \mathfrak{O}[x]$ in the category *Ont* of small ontologies so that, for every ontology $\mathfrak{O}'$, every map $\varphi : \mathfrak{O} \longrightarrow \mathfrak{O}'$ in *Ont*, and every entity $\alpha \in \varphi(A)$ in $\mathfrak{O}'$, there exists a unique map $\varphi' : \mathfrak{O}[x] \longrightarrow \mathfrak{O}'$ such that $\varphi'\eta_x = \varphi$ and $\varphi'(x) = \alpha$. We have not spelled out the definition of "map" in *Ont*.

One way of constructing $\mathfrak{G}[x]$ is to adjoin a new entity $x \in A$ and make sure that expressions involving this new entity still satisfy the equations of an ontology. Of course, this could be subsumed under a general construction for partial algebras. Here we shall make this construction precise by first describing a formal system $L(\mathfrak{G}, x)$.

The *formulas* (or terms) of $L(\mathfrak{G}, x)$ are just the types of $\mathfrak{G}$.

The *proofs* of $L(\mathfrak{G}, x)$ are formal expressions (polynomials) defined inductively as follows:

(1) An entity b of type B in $\mathfrak{G}$ is said to be a proof of B.

(2) The indeterminate x is said to be a proof of A.

(3) If $g(x)$ is a proof of $C \Leftarrow B$ and $b(x)$ is a proof of B then $(g(x)^{\langle} b(x))$ is a proof of C.

(4) Nothing else is a proof.

The reader familiar with mathematical logic with recognize that we are talking about proofs on the assumption A.

We now return to the construction of $\mathfrak{G}[x]$ and the map $\eta_x : \mathfrak{G} \longrightarrow \mathfrak{G}[x]$.

CONSTRUCTION 2. The *types* of $\mathfrak{G}[x]$ are the terms of $L(\mathfrak{G}, x)$. The *entities* of $\mathfrak{G}[x]$ are equivalence classes of proofs, according to the equivalence relation $\equiv$ to be defined presently.

The types T, $B \Leftarrow C$, and $B \wedge C$ of $\mathfrak{G}[x]$ are the corresponding formulas in $L(\mathfrak{G}, x)$. The entity I_B of $\mathfrak{G}[x]$ is defined to be the equivalence class $[I_B]$ of the proof I_B of $B \Leftarrow B$ in $L(\mathfrak{G}, x)$. Similarly we define the entities K, S, P, Q, R and 0. Finally we put $[f]^{\langle}[b] = [f^{\langle} b]$.

The map $\eta_x : \mathfrak{G} \longrightarrow \mathfrak{G}[x]$ is defined by $\eta_x(B) = B$, $\eta_x(b) = [b]$.

The equivalence relation $\equiv$ is the smallest relation between proofs $b(x)$ and $b'(x)$ of the same formula B in $L(\mathfrak{G}, x)$ which satisfies the reflexive, symmetric, transitive and substitution laws, as well as the following conditions (8)':

(i)' $I^{\langle} b(x) \equiv b(x)$, where $b(x)$ is a proof of B in $L(\mathfrak{G}, x)$,

(ii)' $(K^{\langle} b(x))^{\langle} c(x) \equiv b(x)$,

and so on, until (ix)'. Note that the conditions

(9)' $u_t \equiv v_t \quad (t=1, \ldots, n)$,

do not depend on x, hence are special cases of the reflexive law in view of the equations (9).

6. FUNCTIONAL COMPLETENESS

The so-called Deduction Theorem in logic asserts that if B is provable from the assumption A then $B \Leftarrow A$ is provable without this assumption. In the present context this is expressed as follows.

LEMMA 1. To every proof $b(x)$ of a formula B in $L(\mathfrak{G}, x)$ there corresponds canonically an entity $\lambda_x b(x)$ of type B in $\mathfrak{G}$.

Proof. (1) If $b(x) = b$ is already in $\mathfrak{G}$, we write $\lambda_x b = K^{\langle} b$.

(2) If $b(x) = x$, we write $\lambda_x x = I_A$.

(3) If $b(x) = c(x)^{\langle} d(x)$, we write $\lambda_x b(x) = (S^{\langle} \lambda_x c(x))^{\langle} \lambda_x d(x)$.

It may happen that both (1) and (3) apply, namely when $b(x) = c^{\langle} d$, where c and d are in $\mathfrak{G}$. Then equation (viii) says precisely that the two ways of constructing $\lambda_x b(x)$ coincide.

LEMMA 2. If $a(x)$ is any proof of A, then $\lambda_x b(x)^{\langle} a(x) \equiv b(a(x))$.

Proof. The proof is by induction on the length of $b(x)$. One checks the three cases separately.

(1) $\lambda_x b`a(x) = (K`b)`a(x) = b$, by (ii).

(2) $\lambda_x x`a(x) = I`a(x) = a(x)$, by (i).

(3) $\lambda_x(c(x)`d(x))`a(x) = ((S`\lambda_x c(x))`\lambda_x d(x))`a(x)$

$= (\lambda_x c(x)`a(x))`(\lambda_x d(x)`a(x))$, by (iii),

$= c(a(x))`d(a(x))$,

by inductional assumption.

LEMMA 3. $\lambda_x(c`x) = c$, for any entity c of type $B \Leftarrow A$.

Proof. This is a transcription of equation (ix).

LEMMA 4. If $b(x) \equiv b'(x)$ then $\lambda_x b(x) = \lambda_x b'(x)$.

Proof. We show this by induction on the length of the proof that $b(x) \equiv b'(x)$. If the last step of this proof required the reflexive, symmetric, or transitive law, there is nothing to show. Let us assume it required the substitution law:

$$\frac{c(x) \equiv c'(x) \qquad\qquad d(x) \equiv d'(x)}{c(x)`d(x) \equiv c'(x)`d'(x)}$$

By inductional assumption, we have

$$\lambda_x c(x) = \lambda_x c'(x), \quad \lambda_x d(x) = \lambda_x d'(x),$$

hence

$\lambda_x(c(x)`d(x)) = (S`\lambda_x c(x))`\lambda_x d(x)$

$= (S`\lambda_x c'(x))`\lambda_x d'(x)$

$= \lambda_x(c'(x)`d'(x))$.

It remains to consider the situation where $b(x) \equiv b'(x)$ is an instance of one of the equivalences (i)′ to (vii)′.

For example, take a look at

(i)′ $I^{\langle} b(x) \equiv b(x)$,

where $b(x)$ is any proof of B in $L(\mathfrak{G}, x)$. We want to show that

$$\lambda_x(I^{\langle} b(x)) = \lambda_x b(x).$$

If $b(x) = b$ is an entity of type B in $\mathfrak{G}$, this has the form

$$K^{\langle}(I^{\langle} b) = K^{\langle} b,$$

clearly a consequence of (i). Otherwise, it has the form

(*) $(S^{\langle}(K^{\langle} I))^{\langle} g = g$,

where $g = \lambda_x b(x)$ is of type $A \Leftarrow B$.

We would like (*) to be valid in any ontology, in particular with $g = y$ in the ontology $\mathfrak{G}[y]$, where y is an indeterminate of type $B \Leftarrow A$. (Note that $y = \lambda_x(y^{\langle} x)$ by Lemma 3.) But then we also require the validity of the following:

$$\lambda_y((S^{\langle}(K^{\langle} I))^{\langle} y) = I.$$

In view of Lemma 3, this has the form:

(**) $S^{\langle}(K^{\langle} I) = I$.

Conversely, (**) clearly implies (*).

We adopt (**) as the first equation $u_1 = v_1$ in the set (9).

Similarly, equations (ii) to (ix) lead to the choice of $u_2 = v_2, \ldots, u_n = v_n$. For example, equation (vii) leads to $K^{\langle} 0 = I_T$. We refrain from carrying out any further details here. Such a finite set of equations was first obtained by Rosser, see for example the discussion in the book by Rosenbloom.

If A is a ring, an element of the polynomial ring $A[x]$ can be written uniquely in the form $a_0 + a_1 x + \ldots + a_n x^n$, where the $a_i \in A$. For ontologies the corresponding result is even nicer: every entity in $\mathfrak{G}[x]$ can be written in the form $f^{(}x$, where f is an entity in $\mathfrak{G}$. Here is the precise result:

PROPOSITION 1. Let $\mathfrak{G}$ be an ontology, x an indeterminate of type A, $[b(x)]$ an entity in $\mathfrak{G}[x]$ of type B. Then there exists a unique entity f of type $B \Leftarrow A$ in $\mathfrak{G}$ such that $[b(x)] = [f]^{(}[x]$.

Proof. Taking $f = \lambda_x b(x)$, we see from Lemma 2 that $[b(x)] = [f^{(}x] = [f]^{(}[x]$. Suppose also $[b(x)] = [g]^{(}[x] = [g^{(}x]$, then $f^{(}x \equiv g^{(}x$, hence, by Lemmas 3 and 4,

$$f = \lambda_x(f^{(}x) = \lambda_x(g^{(}x) = g.$$

Another consequence of Lemmas 3 and 4 is the following, which may also be obtained as a corollary to Proposition 1.

PROPOSITION 2. Let $\mathfrak{G}$ be an ontology, x an indeterminate of type A. Then the canonical map $\eta_x : \mathfrak{G} \to \mathfrak{G}[x]$ is one-to-one on entities.

Proof. Suppose $\eta_x b = \eta_x b'$, then $b \equiv b'$, hence $b^{(}x \equiv b'^{(}x$. By Lemmas 3 and 4

$$b = \lambda_x(b^{(}x) = \lambda_x(b'^{(}x) = b'.$$

7. CARTESIAN CLOSED CATEGORIES AND ONTOLOGIES

We wish to investigate the connection between ontologies and cartesian closed categories. First we shall show how to construct a cartesian closed category $(\mathfrak{C}, T, \wedge, \Leftarrow, 0, p, q, \langle\,\rangle, e, {}^*)$ from a given ontology $\mathfrak{G}$.

CONSTRUCTION 3. The objects of $\mathfrak{C}$ are the types of $\mathfrak{G}$. In particular, we thus have an object T and objects $A \wedge B$ and $A \Leftarrow B$ for given objects A and B. In view of the natural isomorphism $[A, B] \cong [T, B \Leftarrow A]$, we could define the maps $A \to B$ as entities of type $B \Leftarrow A$. However, the following is more interesting.

Let us assign to each type A of $\mathfrak{G}$ once and for all an indeterminate x_A. We shall say that the maps $A \longrightarrow B$ in $\mathfrak{C}$ are polynomials of type B in $\mathfrak{G}[x_A]$. In particular, the polynomial $[x_A]$ will be regarded as the identity map on A. Composition of maps is defined by substitution of polynomials. Thus, let $f: A \longrightarrow B$ and $g: B \longrightarrow C$, then $f = [b(x_A)] \in B$ in $\mathfrak{G}[x_A]$ and $g = [c(x_B)] \in C$ in $\mathfrak{G}[x_B]$, and we define

$$gf = [c(b(x_A))] \in C \text{ in } \mathfrak{G}[x_A].$$

Furthermore, we define

$$0(A) = [0] \text{ in } \mathfrak{G}[x_A],$$

$$p(A,B) = [P_{A,B}{}^{\langle} x_{A \wedge B}],$$

$$q(A,B) = [Q_{A,B}{}^{\langle} x_{A \wedge B}],$$

$$\langle [b(x_A)], [b'(x_A)] \rangle = [(R_{B,C}{}^{\langle} b(x_A))^{\langle} c(x_A)],$$

$$e(A,B) = [(P_{A \Leftarrow B, B}{}^{\langle} x_{(A \Leftarrow B) \wedge B})^{\langle} (Q_{A \Leftarrow B, B}{}^{\langle} x_{(A \Leftarrow B) \wedge B})],$$

$$c(x_{A \wedge B})^{*} = [\lambda_{x_B} c(\langle x_A, x_B \rangle)].$$

A routine calculation shows that the equations of a cartesian closed category are satisfied.

Conversely, let there be given a cartesian closed category $(\mathfrak{C}, T, \wedge, \Leftarrow, \ldots)$. We shall try to construct an ontology $\mathfrak{G}$ from it.

CONSTRUCTION 4. The types of $\mathfrak{G}$ are the objects of $\mathfrak{C}$. In particular, we thus have a type T and types $A \wedge B$ and $A \Leftarrow B$ for given types A and B.

The entities in $\mathfrak{G}$ of type A are the maps $T \longrightarrow A$.

If f is of type $A \Leftarrow B$ and b is of type B, we define $f^{\langle} b = e(A,B)\langle f, b \rangle$. It is easily seen that $f^{\langle} b$ can also be obtained by composing $f^{\langle}$ with b.

We define seven families of entities as follows:

$I_A = 1_A{}^{)}$,

$K_{A,B} = (p(A,B)^*)^{)}$,

$S_{A,B,C}$ = see below,

$P_{A,B} = (p(A,B))^*$,

$Q_{A,B} = (q(A,B))^*$,

$R_{A,B} = (1_{A \wedge B}{}^*)^{)}$

$0 = 1_T = 0_T$.

Finally,

$$S_{A,B,C} = (e(A,B)\ \langle u, v\rangle)^{**)},$$

where

$$u = e(A \Leftarrow B, C)\ \langle p(F, B \Leftarrow C)\, p(E, C),\ q(E, C)\rangle,$$

$$v = e(B, C)\ \langle p(E, C)\, q(F, B \Leftarrow C),\ q(E, C)\rangle,$$

$$F = (A \Leftarrow B) \Leftarrow C,$$

$$E = F \wedge (B \Leftarrow C).$$

To prove that this construction works, we must show that, under the above interpretations of I_A, $K_{A,B}$, etc., equations (i) to (ix) and the equations $u_i = v_i$ $(i=1, \ldots, n)$, are in fact valid in any cartesian closed category. Failing this, we should incorporate the missing equations into the definition of an "ontological category". I have checked equations (i) to (ix), but have not yet found the time and inclination to carry out the calculations involving the equations $u_i = v_i$. In view of Szabo's decision procedure, these calculations could, in principle, be carried out by a computer.

In any event, we should obtain an equivalence between the category of (small) ontologies and the category of (small) ontological categories. If luck is with us, all cartesian closed categories are ontological.

8. TYPE THEORETICAL FOUNDATIONS

We have established two connections between cartesian closed categories and logic. On the one hand, we have used the positive intuitionist propositional calculus to construct free cartesian closed categories. On the other hand, we have applied cartesian closed categories to the study of combinatory logic with types.

Further progress is possible in two directions. On one side, we could enrich the intuitionist logic to include negation, disjunction, and quantifiers. On the other side, we could enrich the combinatory logic to make it more useful to the foundation of mathematics.

While pure combinatory logic is nothing more than an analysis of the substitution process in mathematics, it had been intended from the earliest days to enlarge its scope to embrace the foundation of all mathematics. This was already the idea of Schönfinkel in the very first article on the subject. Because combinatory logic was then done without types, these early hopes shattered on Russell's paradox. As we are using types here, it may be fruitful to take another look at this program. Inevitably, we shall follow the footsteps of Lawvere.

As a first step, we could enrich our cartesian closed categories to be also cocartesian. Thus we want to introduce finite coproducts, that is, an initial object F and the coproduct $A \vee B$ of two objects. Of course, these come equipped with maps

$$F \longrightarrow A \qquad \begin{matrix} A \longrightarrow A \vee B \\ B \longrightarrow A \vee B \end{matrix} \qquad \frac{A \longrightarrow C \qquad B \longrightarrow C}{A \vee B \longrightarrow C}$$

satisfying appropriate equations. Free bicartesian closed categories may then be constructed from the intuitionist propositional system involving T, $\wedge$, $\Rightarrow$, F, and $\vee$.

In particular, we have the coproduct $T \vee T = \underline{2}$, usually written $\underline{1} + \underline{1} = \underline{2}$, equipped with maps

$$\underline{t} : T \longrightarrow \underline{2}, \qquad \qquad \frac{T \xrightarrow{a} C \qquad T \xrightarrow{b} C}{\underline{2} \xrightarrow{[a,b]} C}$$

$$\underline{f} : T \longrightarrow \underline{2},$$

subject to the equations

$$[a,b]\underline{t} = a, \qquad [a,b]\underline{f} = b, \qquad [h\underline{t}, h\underline{f}] = h.$$

In the corresponding ontology, we have entities

$$\underline{t} \in \underline{2}, \qquad \underline{f} \in \underline{2}, \qquad \pi \in (C \Leftarrow \underline{2}) \Leftarrow (C \wedge C),$$

where

$$\pi`\langle a,b\rangle = [a,b]^{\prime} = \{a,b\}$$

say, so that

$$\{a,b\}`\underline{t} = a, \qquad \{a,b\}`\underline{f} = b, \qquad \{p`\underline{t},\ p`\underline{f}\} = p.$$

A *proposition* is an entity of type $\underline{2}$, for example, $\underline{t}$ and $\underline{f}$ are propositions, but there may be others. We may regard *negation* as an entity N of type $\underline{2} \Leftarrow \underline{2}$. Since it reverses truth values,

$$N = \{\underline{f}, \underline{t}\}.$$

Conjunction, *disjunction*, *implication*, and *equivalence* are entities of type $(\underline{2} \Leftarrow \underline{2}) \Leftarrow \underline{2}$, in fact, they are

$$\{I,\ K`\underline{f}\}, \quad \{K`\underline{t}, I\}, \quad \{I,\ K`\underline{t}\}, \quad \{I, N\}$$

respectively, as we see from the usual truth tables.

Three-valued logic may be done similarly, if one makes use of the coproduct $T \vee T \vee T$.

A *predicate* for entities of type A is an entity of type $\underline{2} \Leftarrow A$. Quantifiers may be defined in terms of *universality* U. We take U to be a new entity of type $\underline{2} \Leftarrow (\underline{2} \Leftarrow A)$, subject to this postulate:

$$U^{\backprime} p = \underline{t} \quad \text{if and only if} \quad p = K^{\backprime} \underline{t}.$$

If x is an indeterminate of type A and $p(x)$ is of type $\underline{2}$, we may define

$$\forall_x p(x) \;=\; U^{\backprime} \lambda_x p(x).$$

Existential quantification is defined as usual.

Equality for entities of type A may be regarded as an entity δ of type $\underline{2} \Leftarrow (A \wedge A)$. According to Leibnitz, it should be defined so that

$$\delta^{\backprime} \langle a, b \rangle = \forall_x (x^{\backprime} a \Leftrightarrow x^{\backprime} b),$$

where x is an indeterminate of type $\underline{2} \Leftarrow A$. Here $p \Leftrightarrow q$ means $(\{I, N\}^{\backprime} p)^{\backprime} q$. It will be noted that $x^{\backprime} a$ is the proposition which asserts that $a \in x$. One may also obtain the *singleton* operation ι of type $(\underline{2} \Leftarrow A) \Leftarrow A$ so that

$$(\iota^{\backprime} a)^{\backprime} b \;=\; \delta^{\backprime} \langle a, b \rangle.$$

We may write

$$\{x \in A \mid p(x)\} \;=\; \lambda_x p(x),$$

when x is an indeterminate of type A and $p(x)$ has type $\underline{2}$. To get the usual set theory, we should postulate the axiom of extensionality and the axiom of choice. The axiom of infinity is best introduced together with a new type and three new entities, as we shall see.

Natural numbers were introduced into combinatory logic a long time ago. If f has type $A \Leftarrow A$, one defines

$$f^2 = \lambda_x(f^{\,\prime}(f^{\,\prime}x)),$$

where x is an indeterminate of type A. One may then define

$$2 = \lambda_y y^2,$$

where y is an indeterminate of type $A \Leftarrow A$. Then 2 is an entity of type

$$(A \Leftarrow A) \Leftarrow (A \Leftarrow A) = B$$

say. Other natural numbers of type B are defined similarly.

There is no difficulty in defining the successor function, addition, and multiplication for natural numbers. However, there is trouble with exponentiation, if one sticks to types. It is customary to define n^2 like f^2 above, but the 2 in n^2 has type $B \Leftarrow B$ and not type B. While Dana Scott has shown in another talk at this conference that it is quite consistent to demand that $(B \Leftarrow B) \cong B$, we shall get around the difficulty in a different way.

We introduce a new type $\underline{N}$ for the natural numbers, and three new entities:

$$\underline{0} \in \underline{N}, \quad \underline{S} \in \underline{N} \Leftarrow \underline{N}, \quad \underline{R}_A \in ((A \Leftarrow A) \Leftarrow (A \Leftarrow A)) \Leftarrow \underline{N},$$

the number *zero*, the operation *successor*, and the process of *iteration* (repetition). They are subject to the following equations:

$$(\underline{R}_A{}^{\prime}\,\underline{0})^{\prime} f = I_A, \quad (\underline{R}_A{}^{\prime}(\underline{S}^{\prime} n))^{\prime} f = f \circ ((\underline{R}_A{}^{\prime} n)^{\prime} f),$$

where $f \in A \Leftarrow A$, $n \in \underline{N}$, and where

$$f \circ g = \lambda_x(f^{\,\prime}(g^{\,\prime}x)).$$

This is an equational version of the Peano-Lawvere axiom. Primitive recursive function theory can be carried out for ontologies enriched in this manner.

Finally, to bring this paper to a happy conclusion, let us define a *dogma* as a cartesian and cocartesian closed category, with universality and naturals, satisfying the axioms of extensionality and choice. (Dogmas are similar to Lawvere's doctrines, they differ from the toposes of Lawvere and Tierney by being equational.) If *Dog* is the category of small dogmas, the forgetful functor $\mathcal{Dog} \longrightarrow \mathcal{Cat}$ has a left adjoint. We suggest that the free dogma generated by the empty category may serve for the type theoretical foundations of mathematics. (We have not discussed partial recursive functions and the axiom of replacement, but these may also be accounted for.)

REFERENCES

A.Church, The calculi of lambda-conversion, Annals of Math. Studies 6, Princeton University Press, Princeton 1941.

H.B.Curry and R.Feys, Combinatory Logic, Vol.1, North-Holland Publ.Co., Amsterdam 1958.

S.Eilenberg and G.M.Kelly, Closed categories, Proc.Conference Categorical Algebra, La Jolla 1965, pp.421-562, Springer-Verlag, New York 1966.

T.Fox, Combinatory Logic and cartesian closed categories, M.Sc.Thesis, McGill University 1970.

K.Gödel, On formally undecidable propositions, reproduced in J.van Heijenoort, From Frege to Gödel, pp.592-617, Cambridge 1967.

J.Lambek, Deductive Systems and Categories I, Math.Systems Theory 2 (1958), 287-318.

J.Lambek, Deductive Systems and Categories II, Lecture Notes in Mathematics 86, pp.76-122, Springer-Verlag, Berlin 1969.

F.W.Lawvere, A functorial analysis of logical operations, undated manuscript.

F.W.Lawvere, Functorial semantics of elementary theories, abstract, J.Symbolic Logic 31 (1966), 294-295.

F.W.Lawvere, Theories as categories and the completeness theorem, abstract, J.Symbolic Logic 32 (1967), 562.

F.W.Lawvere, Diagonal arguments and cartesian closed categories, Lecture Notes in Mathematics 92, pp.134-145, Springer-Verlag, Berlin 1969.

F.W.Lawvere, Adjointness in foundations, Dialectica 23 (1969), 281-296.

F.W.Lawvere, Equality in hyperdoctrines and comprehension schema as an adjoint functor, Proc.Symp.Pure Math.17, pp.1-14, Amer.Math.Society, Rhode Island 1970.

F.W.Lawvere, Category-valued higher logic, Dialectica, to appear.

P.C.Rosenbloom, The elements of mathematical logic, Dover Publications, New York 1950.

M.Schönfinkel, On the building blocks of mathematical logic, reproduced in J.van Heijenoort, From Frege to Gödel, pp.355-366, Cambridge 1967.

M.E.Szabo, Proof-theoretic investigations in categorical algebra, Ph.D.Thesis, McGill University 1970.

M.E.Szabo, A categorical equivalence of proofs, to appear.

Mc Gill University

THE FORMALIZATION OF BISHOP'S CONSTRUCTIVE MATHEMATICS

by
Nicolas D. Goodman and John Myhill

One can distinguish two traditions in the study of the foundations of mathematics. The non-constructive tradition, represented today by set theory and category theory, has as its goal the systematization and unification of mathematics. Its most characteristic technique is the use of ever greater abstraction, so that more and more of mathematics can be seen as composed of special cases of a few extremely general concepts and theorems. The constructive tradition, with which we shall primarily be concerned here, moves in a different direction. It is motivated not by a desire to find unity in the sundry activities of mathematicians, but rather by a desire to ensure the reliability of those activities. The constructive mathematician is suspicious of abstraction and typically feels most comfortable with the concreteness and immediacy of combinatorial manipulations. The constructive tradition is represented today by intuitionism, finitism, and much of proof theory. These two tendencies in foundational studies are not incompatible. Rather, it is the interaction between them that is likely to lead to the most fruitful development of foundations as a whole. Current examples include the use of infinite proof-figures in proof theory and the use of elementary, rather than higher order, theories in studying categories. Our subject here is a recent development in constructivity which promises to open new avenues for such interaction.

The first systematic attempt to ensure the reliability of classical mathematics was made in the 1920's under the leadership of David Hilbert. The basic idea of Hilbert's program was (1) that certain simple forms of statements do not involve any problematic notions, (2) that there are certain non-problematic forms of reasoning appropriate to those elementary statements, and (3) that other

---

The work of both authors was supported in part by NSF Grant GP 13019.

mathematical statements have in themselves no meaning, but are merely useful devices to facilitate the proofs of elementary statements; further (4) that the customary forms of reasoning, both with the new ("ideal") notions and with the old ("elementary" or "real") ones, are justified only insofar as it can be shown, by elementary means, that no "real" statements can be obtained by the use of "ideal" ones that could not have been obtained in a purely elementary way. Thus the system of "ideal" mathematics must be (provably) a <u>conservative extension</u> of the system of "real" mathematics.

On some interpretations this conception of mathematics is demonstrably false; i.e., there are certain senses of "real statement" and "ideal statement," perhaps including those which come to mind most naturally, such that the system of ideal statements is <u>not</u> a conservative extension of the system of real statements. For example, if the real statements are (as Hilbert thought they were) of the form $p(x) = 0$ with primitive recursive p, and if ideal statements permit quantification over natural numbers and even the intuitionistic connectives, then (after Gödel) it is easy to see that the system of real statements (Heyting arithmetic HA) is not a conservative extension of the system of real statements (primitive recursive arithmetic). This is true <u>a fortiori</u>, of course, if the system of ideal statements includes variables of higher type, even when these are predicatively interpreted. The prospect for a sensible application of Hilbert's idea to ordinary mathematics seems dismal indeed, if one approaches it in <u>this</u> way.

Nevertheless, Hilbert's program is not dead. One form in which it continues to be actively pursued has been most forcefully propounded by Kreisel (for example, in [7]). The leading idea behind Kreisel's form of Hilbert's program seems to be that the original program failed because it took "reality" in too narrow a sense. The ultimate goal is to find a theory T, possibly incompatible with classical mathematics, which will be evidently constructive and therefore reliable, but which will be sufficiently strong that we can prove, for example, that classical analysis is a conservative extension of that part of T which it has in common with classical analysis. Thus for Kreisel a crucial part of the study of foundations is the search for ever stronger, and therefore more abstract, constructive principles. As this search is carried forward, however, the principles proposed are receding more and more from that combinatorial reality which lies at the heart of mathematics. Indeed, the theory of iterated generalized inductive definitions, as in Tait [11], or the theory of the thinking subject, as in Kreisel [8], seem just as abstract and problematical as the second-order arithmetic for which they are ultimately intended to provide a foundation. Thus even if this program were to succeed, it seems unlikely that it would provide the elimination of the ideal from mathematics which Hilbert had in mind.

We would like to suggest a quite different approach to reviving Hilbert's program, based on the work of Bishop, which involves no great extension of what is considered real, but rather a reformulation of the desired solution by giving up the image of classical mathematics as a formalized theory. Mathematics as practiced, after all, is not a formal game with symbols. What is interesting about a mathematical argument is rarely the system in which it is or could be formalized, but rather the ideas involved. At present, these ideas are typically presented in ideal language. Thus if we simply formalize what mathematicians say they are doing, we are bound to end up with a system whose intended interpretation is thoroughly ideal and hence incomprehensible in "real" terms. On the other hand, if it is actually true that the basic subject-matter of mathematics is real, then it ought to be possible to understand the ideas involved in ordinary mathematics sufficiently well to be able to exhibit their concrete content. In other words, it seems likely that the ideal component of mathematics is merely a superstructure which, in any particular case, can be seen to be inessential. Thus an alternative proposal for the resurrection of Hilbert's program would be to give up thinking of mathematics as codified in a formal system, such as Zermelo-Fraenkel, and instead to think of it as a body of ideas whose real content is to be found piecemeal by direct analysis of those ideas. If such a program were successful, it would lead not to the resolution of the tension between the real and ideal components of mathematical practice, but rather to a more fruitful interaction between them. Abstraction would continue to unify and guide mathematics, but would be held in check by its relation to the combinatorial realities which must ultimately provide its content.

The first systematic attempt to carry out such a program was made by Errett Bishop in [2]. It is only fair to warn the reader that we do not always agree with what Bishop says about what he is doing, and that we have no guarantee that he would agree with what we are saying about what he has done. In his book Bishop gives an account of all of basic classical and abstract analysis in such a way that the concrete content of the theory is beautifully exhibited. The theorems are sometimes stated in a rather different form than we are used to, but the ideas are still clearly recognizable, the definitions and proofs are constructive, and the relation of the theory to concrete combinatorial problems is always evident. Indeed, the theorems in Bishop's form often give more numerical information than in their traditional form. Thus Bishop's work shows that our program can be carried out at least for most of analysis.

It is important to ask what methods Bishop is using in order to carry out his reformulation of analysis. It is all very well to say that his mathematics is "constructive." We would like to have more than that. Specifically, if the reformulation is fully to satisfy our program, then it is desirable that it be carried out in a fragment of classical mathematics which is no stronger than HA. The requirement that we work within classical mathematics is imposed by the consideration that what we are trying to show is precisely that the ideal component of mathematics is ultimately unnecessary. To introduce notions which are classically incomprehensible, like Brouwer's free-choice sequences, or axioms which are classically false, like the assertion that all functions are recursive, is to begin the development of a new mathematics, not to give an analysis of the ideas involved in the mathematics that is actually done. The requirement that the techniques used not be stronger than those codifiable in HA is simply the requirement that the reformulation be carried out using only finitary means. It is also desirable that the theory used be at least formally constructive, in the sense that it have the following E-property: If $\forall x \mathfrak{A}(x)$ is provable, then there is a term t of the theory such that $\mathfrak{A}(t)$ is also provable.

Thus in order to see whether Bishop has actually done what we would like him to have done, we must attempt to formalize the theory in [2] and then see whether the resulting system meets our criteria. Bishop's theory is concerned with the hierarchy of finite types over the natural numbers. Thus we find the natural numbers, real numbers interpreted as certain functions from natural numbers to natural numbers, functions from real numbers to real numbers, functionals from functions of a real variable to functions of a real variable, and so on. We also find various kinds of ordered pairs of objects which themselves fit into this hierarchy. Thus a natural way to describe Bishop's domain of discourse is by the following type structure. We give an inductive definition of the set of (finite) types. First, 0 is a type. Intuitively, 0 is the ground type, or type of the natural numbers. If σ and τ are types, then so are $(\sigma \times \tau)$ and $(\sigma \to \tau)$. Intuitively, $(\sigma \times \tau)$ is the type of pairs whose first component is of type σ and whose second component is of type τ. Also, $(\sigma \to \tau)$ is the type of total functions from objects of type σ to objects of type τ. These are all the types.

In order to refer to objects of this type structure we will need variables $x^\tau, y^\tau, z^\tau, \ldots$ of every type τ, and certain constants of specific types to denote particular objects of those types. Then the terms of our theory are defined inductively. First, each variable of type τ and each constant of type τ is a term of type τ. Second, if t is a term of type $\sigma \to \tau$ and s is a term of type σ, then $t(s)$ is a term of type τ. Third, if t is a term of type τ and x is a variable of type σ, then $(\lambda x.\ t)$ is a term of type $\sigma \to \tau$. These are the only terms. Intuitively, $t(s)$ is the value of t applied to s. Moreover, $(\lambda x^\sigma.\ t)$ is that function of type $\sigma \to \tau$ whose value, for any x of type σ, is t.

As constants of our theory we choose the following:

1) A constant $\overline{0}$ of type 0 to denote the natural number zero.

2) A constant S of type $0 \to 0$ to denote the successor function on the natural numbers.

3) For any types σ and τ, a constant D of type $\sigma \to (\tau \to \sigma \times \tau)$ such that, if x is of type σ and y is of type τ, then $D(x,y)$ is the ordered pair of x and y.

4) For any types σ and τ, constants D_1 of type $\sigma \times \tau \to \sigma$ and D_2 of type $\sigma \times \tau \to \tau$ to denote the inverse pairing functions.

5) For any type σ, a constant R of type

$$(\sigma \to (0 \to \sigma)) \to (\sigma \to (0 \to \sigma))$$

to denote the primitive recursion functional for type σ.

Next we define the formulas of our theory inductively. If s and t are terms of the same type τ, then $s = t$ is a formula. If $\mathfrak{A}$ and $\mathfrak{B}$ are formulas, then so are $\mathfrak{A} \vee \mathfrak{B}$, $\mathfrak{A} \wedge \mathfrak{B}$, and $\mathfrak{A} \to \mathfrak{B}$. If $\mathfrak{A}$ is a formula, and x is any variable, then $\bigvee x\mathfrak{A}$ and $\bigwedge x\mathfrak{A}$ are formulas. These are the only formulas. In writing formulas, we employ the usual conventions. In particular, $\neg\mathfrak{A}$ is $\mathfrak{A} \to S(\overline{0}) = \overline{0}$.

The first theory couched in this language that we wish to describe is HA^{ω} (Heyting arithmetic of all finite types). The underlying logic of HA^{ω} is the many-sorted intuitionistic predicate calculus with identity. We do not assume in HA^{ω} that identity is decidable, as intensional identity might be expected to be, or, on the other hand, that it satisfies any principle of extensionality. The non-logical axioms of HA^{ω} are as follows, where the variables are of suitable types:

1) $S(x) \neq \overline{0}$.

2) $S(x) = S(y) \to x = y$.

3) $D_1(D(x,y)) = x$.

4) $D_2(D(x,y)) = y$.

5) $D(D_1 x, D_2 x) = x$.

6) $(\lambda x.\, t)(x) = t$.

7) $R(x,y,\overline{0}) = y$.

8) $R(x,y,S(z)) = y(R(x,y,z),z)$.

9) From $\mathfrak{A}(\overline{0})$ and $\mathfrak{A}(x) \to \mathfrak{A}(S(x))$, infer $\mathfrak{A}(x)$.

It is evident that HA^{ω} is an extension of HA. Using the technique of Tait [10], we can interpret the variables of HA^{ω} as ranging over the closed terms of HA^{ω} and thus see that HA^{ω} is a conservative extension of HA. The same argument, however, shows that HA^{ω} does not make essential use of the objects of higher type. For, the argument shows that if we add the law of excluded middle to HA^{ω}, thus obtaining a theory with classical logic, the result is a conservative extension of classical arithmetic $\mathbf{Z}$, and therefore still a conservative extension of HA for Π^0_2 - sentences. The problem is that HA^{ω} contains no comprehension principles which could enable us even classically to prove the existence of any functional not explicitly definable. Indeed, for this very reason, HA^{ω} is inadequate for the formalization of most of Bishop's theory in [2].

From an intuitionistic point of view, the natural comprehension principles to add are axioms of choice. We consider three of these. First of all, the <u>axiom of choice</u> at types σ,τ is the schema

$$AC_{\sigma,\tau}\colon\ \Lambda x^{\sigma}\, Vy^{\tau}\mathfrak{A}(x,y) \to Vz^{\sigma\to\tau}\Lambda x^{\sigma}\,\mathfrak{A}(x,z(x)).$$

We write AC for the schema $AC_{\sigma,\tau}$ for all types σ and τ. The <u>axiom of dependent choices</u> at type σ is the schema

$$DC_{\sigma}\colon\ \Lambda x^{\sigma}\, Vy^{\sigma}\mathfrak{A}(x,y) \to \Lambda x^{\sigma}\, Vf^{0\to\sigma}[f(\overline{0}) = x \wedge \Lambda z^{0}\,\mathfrak{A}(f(z),\ f(S(z)))].$$

We write DC for the schema DC_{σ} for all types σ. Finally, the <u>relativized axiom of dependent choices</u> at type σ is the schema

$$\mathrm{RDC}_\sigma: \quad \Lambda x^\sigma[\mathfrak{A}(x) \to \mathrm{V} y^\sigma(\mathfrak{A}(y) \wedge \mathfrak{B}(x,y))]$$

$$\to \Lambda x^\sigma[\mathfrak{A}(x) \to \mathrm{V} f^{0\to\sigma}[f(\bar{0}) = x \wedge \Lambda z^0 \mathfrak{B}(f(z), f(S(z)))]].$$

Again, we write RDC for the schema RDC_σ for all types σ. Obviously RDC_σ implies DC_σ. As was pointed out to Goodman by Peter Freyd, in the presence of the primitive recursion functional R, the axiom $\mathrm{AC}_{\sigma,\sigma}$ also clearly implies DC_σ. We do not know whether either AC or RDC implies the other.

For convenience, let S denote the theory HA^ω + AC + RDC. In this theory the higher types are not trivial. Indeed, the theory obtained from S by adding excluded middle is as strong as full classical analysis. Moreover, S is strong enough to formalize everything in Bishop [2] except possibly the theory of Borel sets. Now, the remarkable fact is the

<u>Theorem</u> (Goodman): S is a conservative extension of HA.

The proof requires a detailed analysis of the logic involved in S. (For one version of the argument, see Goodman [5]. An improved version is in Goodman [6].) Moreover, it follows from Theorem 7.2 of Troelstra [13] that S has the E-property. On the basis of these results, and of the conjectured formalizability of Bishop's work in S, we may say that Bishop has actually carried out the desired finitary reformulation of classical analysis.

Let us look more carefully at our porposed formalization of Bishop's constructive analysis. First of all, although Bishop never uses it, one would like to have an <u>axiom of extensionality</u> in the following form:

$$\mathrm{Ext}_{\sigma,\tau}: \quad \Lambda y^{\sigma\to\tau}\,\Lambda z^{\sigma\to\tau}[\Lambda x^\sigma(y(x) = z(x)) \to y = z].$$

Let Ext be the schema $\mathrm{Ext}_{\sigma,\tau}$ for all σ and τ. Then we conjecture that S + Ext is still a conservative extension of HA. The reason that this is not immediately clear is that Ext makes AC less obvious from an intuitionistic point of view. To know that for every x there is a y is just to have a procedure which, applied to any x, gives a suitable y. However, there is no *a priori* guarantee that this procedure depends only extensionally on x.

There is a much more fundamental problem involved in carrying out a formalization of Bishop's work in a theory like S. That is the problem of what to do about references to sets. Such references occur repeatedly in Bishop [2]. An old slogan has it that a set exists just in case we have a definition of it. (See Bishop [2], p. 2.) If we are to take this seriously, then it is natural to interpret Bishop as having in mind only a *virtual* theory of sets. That is, we simultaneously introduce notation $\{\cdot|\cdot\}$ and the symbol $\in$ by means of the contextual definition

$$t^{\tau} \in \{x^{\tau}|\mathfrak{A}(x^{\tau})\} \leftrightarrow \mathfrak{A}(t^{\tau}).$$

For convenience, we may also introduce free variables X^{τ}, Y^{τ}, Z^{τ}, ... ranging over such formal abstracts $\{x^{\tau}|\mathfrak{A}(x^{\tau})\}$. Then it is clear that if we have proved $\mathfrak{B}(X^{\tau})$, then we could also have proved $\mathfrak{B}(\{x^{\tau}|\mathfrak{A}(x^{\tau})\})$. Thus this theory, which we may call S^*, is a conservative extension of S. The theory S^* suffices to formalize most of Bishop's references to sets.

The theory S^* does not, however, eliminate all references to sets. The problem is that Bishop occasionally wants to quantify over sets. The obvious first idea is to add a new type (τ) for each type τ to be the type of species of objects of type τ. We then add quantification over these new types together with a comprehension principle of the following form:

$$\bigvee X^{(\tau)} \bigwedge x^{\tau}(x \in X \leftrightarrow \mathfrak{A}(x)] ,$$

where X is not free in $\mathfrak{A}(x)$. As a matter of fact, Bishop himself now favors such a course. Unfortunately, this extended theory no longer has the E-property. (Myhill intends to publish this result separately.) What is more serious is that a negative interpretation, as in Gödel [4], shows that this theory is proof-theoretically no weaker than classical analysis. (See Myhill [9].) Actually, the extension of the negative interpretation from arithmetic to type theory is not quite straightforward, since besides the negative translation we have also to restrict the quantifiers to stable sets. In fact the definition of 'stable set' given in [9] is wrong: it should be defined simultaneously with an equivalence relation $\underset{S}{\equiv}$, as follows:

$$S(x_0) \equiv x = x$$

$$x_0 \underset{S}{\equiv} y_0 \equiv x = y$$

$$S(x_{i+1}) \equiv \Lambda y_i [S(y) \wedge \neg\neg y \in x \rightarrow y \in x]$$

$$\wedge \Lambda y_i \; \Lambda z_i [S(y) \wedge S(z) \wedge y \underset{S}{\equiv} z \wedge y \in x \rightarrow z \in x]$$

$$x_{i+1} \underset{S}{\equiv} y_{i+1} \equiv \Lambda z_i [S(z) \rightarrow (z \in x \leftrightarrow z \in y)] .$$

Here $\rightarrow$ and $\leftrightarrow$ are the intuitionistic connectives, and the theorem is:
If $\mathfrak{A}$ is closed and provable in the classical system and $\mathfrak{A}_S$ is formed by restricting all variables in the negative translation of $\mathfrak{A}$ to S, then $\mathfrak{A}_S$ is provable in the intuitionistic system. From these two results of Myhill we see that to follow this proposal would be to vitiate the whole program. Using intuitionistic logic does not guarantee that one's mathematics has computational content.

In [3], Bishop suggests a second approach to this problem which is more satisfactory. Let us extend S by adding two logical principles which, though classically valid, are not in general provable in S. The first is a generalized form of Markov's principle at type τ:

$$MP_\tau: \; \neg \Lambda x^\tau \mathfrak{A}(x) \rightarrow V x^\tau \neg \mathfrak{A}(x),$$

where $\mathfrak{A}(x)$ is quantifier-free. The second, sometimes called the implication principle at type τ, is

$$IP_\tau: \; [\Lambda x^\sigma \mathfrak{A}(x) \rightarrow V y^\tau \mathfrak{B}(y)] \rightarrow V y^\tau [\Lambda x^\sigma \mathfrak{A}(x) \rightarrow \mathfrak{B}(y)],$$

where $\mathfrak{A}(x)$ is again quantifier-free. Then it turns out that for every formula $\mathfrak{A}(x^\tau)$ there is a term t of some type $\rho \to (\sigma \to (\tau \to 0))$ such that

$$S + MP + IP \vdash \mathfrak{A}(x^\tau) \leftrightarrow \bigvee y^\rho \bigwedge z^\sigma (txyz = 0) .$$

This result is due to Yasugi [14]. From this it is straightforward, using the Gödel interpretation, to obtain the E-property for $S + MP + IP$. Troelstra's proof of the E-property for S is a very significant improvement since it does not involve the use of an ad hoc normal form for formulas. The theory $S + MP + IP$ is still a conservative extension of HA for Π_2^0-sentences. But in this theory we can evidently quantify over a doubly ramified hierarchy of sets of objects of type τ by defining

$$x^\tau \in F^{\rho\to(\sigma\to(\tau\to 0))} \leftrightarrow \bigvee y^\rho \bigwedge z^\sigma (Fxyz = 0) .$$

The main difficulty is that the ramification by the type of F is unnatural. From a constructive point of view, moreover, there is no reason to accept the principles MP or IP.

A third alternative, which would suffice for all the applications of set quantification that Bishop needs, and which has the merit of making clear constructive sense, would be to extend S by adding a suitable generalized inductive definition. Perhaps the simplest form this could take would be the addition of Kreisel's class K of constructive ordinals, as described, for example, in Troelstra [12]. Specifically, we add a new primitive predicate K which can be meaningfully applied to objects of type $0 \to 0$. For the moment, let us write n, m, p for x^0, y^0, z^0 and f, g, h for $x^{0\to 0}$, $y^{0\to 0}$, $z^{0\to 0}$. We suppose we are given a primitive recursive one-to-one mapping of the finite sequences of natural numbers onto the natural numbers in such a way that 0 corresponds to the empty sequence. We write $< n >$ for the one-term sequence whose single term is n, and write $n*m$ for the concatenation of the sequences n and m. Then, following Troelstra, the new axioms are as follows:

1. $\bigwedge n K(\lambda m.\ S(n))$

2. $\bigwedge f[f(0) = 0 \wedge \bigwedge n K(\lambda m.\ f(<n> * m)) \rightarrow K(f)]$

3. $\bigwedge n \mathfrak{A}(\lambda m.\ S(n)) \wedge \bigwedge f[f(0) = 0 \wedge \bigwedge n \mathfrak{A}(\lambda m.\ f(<n> * m)) \rightarrow \mathfrak{A}(f)] \rightarrow$
$\bigwedge g[K(g) \rightarrow \mathfrak{A}(g)]$.

On this basis, the class of Borel sets can be defined explicitly in terms of K. The disadvantage of this approach is that it would constitute an essential strengthening of the theory, taking us beyond what anyone would call finitistic, or even predicative.

In [3], Bishop makes a fourth suggestion, which we believe to be by far the best. Let us consider a specific context in which a mathematician might actually need to quantify over sets. Suppose we are studying the topology of the real line, and need to be able to quantify over the open subsets of the line. Now, an arbitrary open set of reals is a countable union of rational intervals. A rational interval, in turn, can be coded by a natural number. Thus we can index the open sets by sequences f of natural numbers, where a real number x is in f just in case there is an n such that x is in the rational interval coded by f(n). In this way, quantification over open sets is eliminated in favor of quantification over sequences without the need for any extension of S. This is done, however, not in a routine way, but by using information provided by the classical theory about the sets in question. The same sort of elimination is sketched by Bishop in [3] for the much more difficult case of the Borel sets. (Some of the details are in Beeson [1].) Thus one may hope that the ultimate bastion of classical idealism, set theory, can be made to give way piecemeal to the insights which, in particular cases, it gives into the structure of its own objects.

REFERENCES

1. M. J. Beeson, Metamathematics of constructive analysis, Ph.D. Thesis, Stanford University, 1971.

2. E. Bishop, Foundations of Constructive Analysis (McGraw-Hill, New York, 1967).

3. E. Bishop, Mathematics as a numerical language, in: Intuitionism and Proof Theory, ed. J. Myhill, A. Kino, and R. E. Vesley (North-Holland, Amsterdam, 1970) 53-71.

4. K. Gödel, On intuitionistic arithmetic and number theory, trans. M. Davis, in: The Undecidable, ed. M. Davis (Raven Press, Hewlett, N.Y., 1965) 75-81.

5. N. D. Goodman, Intuitionistic arithmetic as a theory of constructions, Ph.D. Thesis, Stanford University, 1968.

6. N. D. Goodman, The theory of the Gödel functionals, In preparation.

7. G. Kreisel, Hilbert's programme, in: Philosophy of Mathematics, ed. P. Benaceraff and H. Putnam (Prentice-Hall, Englewood Cliffs, N.J., 1964) 157-180.

8. G. Kreisel, Informal rigour and completeness proofs, in: Problems in the Philosophy of Mathematics, ed. I. Lakatos (North-Holland, Amsterdam, 1967) 138-171.

9. J. Myhill, Embedding classical type theory in 'intuitionistic' type theory, in: Axiomatic Set Theory, ed. D. S. Scott (Proc. Symp. Pure Math. 13, Part I, 1971) 267-270.

10. W. W. Tait, Intensional interpretation of functionals of finite type I, J. Symb. Log. 32 (1967) 198-212.

11. W. W. Tait, Constructive reasoning, in: Logic, Methodology and Philosophy of Science III, ed. B. van Rootselaar and J. F. Staal (North-Holland, Amsterdam, 1968) 185-199.

12. A. S. Troelstra, The theory of choice sequences, in: Logic Methodology and Philosophy of Science III, ed. B. van Rootselaar and J. F. Staal (North-Holland, Amsterdam, 1968) 201-223.

13. A. S. Troelstra, Notions of realizability for intuitionistic arithmetic and intuitionistic arithmetic in all finite types. To appear.

14. M. Yasugi, Intuitionistic analysis and Gödel's interpretation, J. Math. Soc. Japan, 15 (1963) 101-112.

STATE UNIVERSITY OF NEW YORK AT BUFFALO

Continuous Lattices

BY

Dana Scott

ABSTRACT

Starting from the topological point of view a certain wide class of T_0-spaces is introduced having a very strong extension property for continuous functions with values in these spaces. It is then shown that all such spaces are complete lattices whose lattice structure determines the topology - these are the *continuous lattices* - and every such lattice has the extension property. With this foundation the lattices are studied in detail with respect to projections, subspaces, embeddings, and constructions such as products, sums, function spaces, and inverse limits. The main result of the paper is a proof that every topological space can be embedded in a continuous lattice which is homeomorphic (and isomorphic) to its own function space. The function algebra of such spaces provides mathematical models for the Church-Curry λ-calculus.

CONTENTS

0. Introduction

Through a roundabout chain of mathematical events I have become interested in T_0-spaces, those topological spaces satisfying the weakest separation axiom to the effect that two distinct points cannot share the same system of open neighborhoods. These spaces seem to have been originally suggested by Kolmogoroff and were introduced first in Alexandroff and Hopf (1935). Subsequent topology textbooks have dutifully recorded the definition but without much enthusiasm: mainly the idea is introduced to provide exercises. In the book Čech (1966) for example, T_0-spaces are called *feebly semi-separated* spaces, which surely is a term expressing mild contempt. Some interest has been shown in finite T_0-spaces (finite T_1-spaces are necessarily discrete), but generally topology seems to go better under at least the Hausdorff separation axiom. The reason for this is no doubt the strong motivation we get from geometry, where points *are* points and where distinct points *can* be separated.

What I hope to show in this paper is that from a less geometric point of view T_0-spaces can be not only interesting but also natural. The interest for me lies in the construction of *function spaces*, and the main result is the production of a large number of T_0-spaces D such that D and $[D \to D]$ are *homeomorphic*. Here $[D \to D]$ is the space of all continuous functions from D into D with the topology of point-wise convergence (the product topology). It will be shown that every space can be embedded in such a space D, and that D can be chosen to have quite strong extension properties for D-valued continuous functions. These properties make D most convenient for applications to logic and recursive function theory, which was the author's original motivation. Some of the facts about these spaces seem to be most easily proved with the aid of some lattice theory, a circumstance that throws new light on the connections between topology and lattices. In fact, the required spaces are at the same time complete lattices whose topology is determined by the lattice structure in a special way, whence my title.

1. INJECTIVE SPACES. All spaces are T_0-spaces, and we begin by defining a class of spaces to be called *injective*.

1.1 Definition. A T_0-space D is *injective* iff for arbitrary spaces X and Y if $X \subseteq Y$ as a subspace, then every continuous function $f: X \to D$ can be extended to a continuous function $\bar{f}: Y \to D$. As a diagram we have:

$$\begin{array}{ccc} X & \subseteq & Y \\ & {\scriptstyle f}\searrow \quad \swarrow{\scriptstyle \bar{f}} & \\ & D. & \end{array}$$

Some people will object to this terminology because I use the subspace relationship rather than a monomorphism in the category of T_0-spaces and continuous maps. However, only the trivial 1-point space is injective in the sense of monomorphisms in that category, and so the notion is uninteresting. If the reader prefers another terminology, I do not mind. As we shall see these spaces have very strong retraction properties.

A slightly less trivial example of an injective space is the 2-point space $\mathbb{O}$ with "points" $\bot$ and $\top$ where $\{\top\}$ is open but $\{\bot\}$ is not. (This space is sometimes called the *Sierpinski Space*.)

1.2 Proposition. *The space $\mathbb{O}$ is injective.*

Proof: As is obvious, the continuous maps $f: X \to \mathbb{O}$ are in a one-one correspondence with the open subsets of X (consider $f^{-1}(\{\top\})$). If $X \subseteq Y$ as a subspace, then an open subset of X is the restriction of some open subset of Y. Thus any $f: X \to \mathbb{O}$ can be extended to $\bar{f}: Y \to \mathbb{O}$. □

1.3 Proposition. *The Cartesian product of any number of injective spaces is injective under the product topology.*

Proof: The argument is standard. A map into the product can be projected onto each of the factors. Each of these projections can be extended. Then the separate maps can be put together again to make the required extended map into the product. □

We now have a large number of injective spaces, and further examples could be found using the next fact.

1.4 Proposition. *A retract of an injective space is injective.*

Proof: Let D be injective. By a retract of D we understand a subspace $D' \subseteq D$ for which there exists a retraction map $j: D \to D'$ such that

$$D' = \{x \in D : j(x) = x\}.$$

Then if $f: X \to D'$ and $X \subseteq Y$, we have $f: X \to D$ as a continuous map also. Taking $\overline{f}: Y \to D$, we have only to form

$$j \circ \overline{f} : Y \to D'$$

to show that D' is injective. □

The relationship between arbitrary T_0-spaces and the injective spaces is given by the embedding theorem.

1.5 Proposition. *Every T_0-space can be embedded in an injective space; in fact, in a Cartesian power of the 2-element space $\mathbb{O}$.*

Proof: The proof is well known (cf. Čech (1966), Theorem 26B.9, p. 484.) But we give the argument for completeness sake. Let X be the given space, and let $\mathfrak{O}$ be the class of open subsets of X. Let

$$D = \mathbb{O}^{\mathfrak{O}}$$

be the Cartesian power of $\mathbb{O}$. Then D is injective by 1.3. Define the map $e: X \to D$ by:

$$e(x)(U) = \begin{cases} \top & \text{if } x \in U, \\ \bot & \text{if } x \notin U, \end{cases}$$

for $x \in X$ and $U \in \mathfrak{O}$. This map e is *continuous* in view of the topology given to $\mathbb{O}$ and to D. The map e is one-one, because X is T_0. Finally, if $U \subseteq X$ is open, then

$$\begin{aligned} e(U) &= \{e(x) : x \in U\} \\ &= \{e(x) : e(x)(U) = \top\} \\ &= e(X) \cap \{t \in D : t(U) \in \{\top\}\}, \end{aligned}$$

which shows that the image $e(U)$ is open in the subspace $e(X) \subseteq D$. Therefore $e: X \to D$ is an embedding of X as a subspace in D. □

1.6 Corollary. *The injective spaces are exactly the retracts of the Cartesian powers of $\mathbb{O}$.*

Proof: Such a retract is injective by 1.4. If D is injective, then it is (homeomorphic to) a subspace of a power of $\mathbb{O}$. But since D is injective the identity function on the subspace to itself can be extended to the whole of the power of $\mathbb{O}$ providing the required retraction. □

1.7 Corollary. *A space is injective iff it is a retract of every space of which it is a subspace.*

Proof: As in the proof of 1.6, this property is obvious for injective spaces. But in view of 1.5 every such space is a retract of a power of $\mathbb{O}$ and hence is injective. □

As a result of these very elementary considerations, the injective space could be called *absolute retracts*, if one remembers to modify the standard definitions by using *arbitrary* subspaces rather than just *closed* subspaces. Note too that it is easy to show that the only continuous maps $e:X \to Y$ for which the extension property

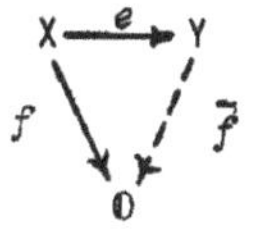

could hold for *all* continuous $f:X \to \mathbb{O}$ *are* embeddings as subspaces. Thus it would seem that we have a reasonably good initial grasp of the notion of injective spaces, but further constructions are considerably facilitated by the introduction of the lattice structure.

2. CONTINUOUS LATTICES. Every T_0-space becomes a partially ordered set under the definition:

$$x \sqsubseteq y \text{ iff whenever } x \in U \text{ and } U \text{ is open,}$$
$$\text{then } y \in U.$$

Indeed, though this relation is reflexive and transitive, the condition that it be *antisymmetric* is exactly equivalent to the T_0-axiom.

In the converse direction, every partially ordered set $\langle X, \sqsubseteq \rangle$ can be so obtained, for we have only to define $U \subseteq X$ as being open if it satisfies the condition:

$$(i) \text{ whenever } x \in U \text{ and } x \sqsubseteq y, \text{ then } y \in U.$$

The axioms for partial order make X a T_0-space, because for any $y \in X$ the set

$$\{x \in X : x \not\sqsubseteq y\}$$

is open. This connection is not very interesting, however.

What *is* interesting in topological spaces is *convergence* and the properties of *limit points*. We shall discuss limits in terms of *nets*, in particular in terms of *monotone* nets. A monotone net in a T_0-space X is a function

$$x : I \to X$$

where $\langle I, < \rangle$ is a directed set and where $i < j$ implies $x_i \sqsubseteq x_j$ for all $i,j \in I$. In a T_1-space a monotone net is constant (hence, uninteresting) because the $\sqsubseteq$-relation is the identity. As usual (cf. Kelley (1955), p.66) we say that a net x *converges* to an element y iff whenever U is open and $y \in U$, then for some $i \in I$ we have $x_j \in U$ for all $j > i$. Note that a monotone net x converges to each of its terms x_i. Suppose that a monotone net x converges to an element y which is an *upper bound* to all the terms of x. Then y must be the *least* upper bound, which we write as:

$$y = \bigsqcup \{x_i : i \in I\}$$

To see this, assume that z is any other upper bound with $x_i \sqsubseteq z$ for all $i \in I$. If U is open and $y \in U$, then $x_i \in U$ for some $i \in I$. But then $z \in U$, and so $y \sqsubseteq z$ follows.

We shall find that most of the facts about the topology of the spaces we are concerned with here can be expressed in terms of least upper bounds (lubs). It is not always the case, however, that lubs are limits. Thus, for a partially ordered set X, we impose a further restriction on its topology beyond condition (*i*) for saying when a subset U is open:

(*ii*) whenever $S \subseteq X$ is directed, $\bigsqcup S$ exists, and $\bigsqcup S \in U$, then $S \cap U \neq \emptyset$.

By a directed subset of X we of course mean that it is directed in the sense of the partial ordering $\sqsubseteq$. Note that in this paper directed sets are always *non-empty*. The sets satisfying (*i*) and (*ii*) form the *induced topology* on a partially ordered set X, which is still a T_0-space because the sets

$$\{x \in X : x \not\sqsubseteq y\}$$

remain *open* even in the sense of (*ii*). Obviously a directed set $S \subseteq X$ can be regarded as a net, and now in view of (*ii*) it follows that S converges to $\bigsqcup S$ -- if this lub exists. We can summarize this discussion as follows.

<u>2.1 Proposition</u>. *In a partially ordered set* X *with the induced topology, a monotone net* $x : I \to X$ *with a least upper bound converges to an element* $y \in X$ *iff*

$$y \sqsubseteq \bigsqcup \{x_i : i \in I\}. \quad \Box$$

Our main interest will lie with those partially ordered sets in which *every* subset has a lub: namely, *complete lattices*. If D is such a space we write $\bot = \bigsqcup \emptyset$ and $\top = \bigsqcup D$ for the smallest and largest elements (read: *bottom* and *top*). As is well known, *greatest lower bounds* must exist, for:

$$\bigsqcap S = \bigsqcup \{x \in D : x \sqsubseteq y \text{ for all } y \in S\}$$

gives the definition.

Given a complete lattice D we define

$$x \prec y \text{ iff } y \in \text{Int } \{z \in D : x \sqsubseteq z\},$$

where the interior is taken in the sense of the induced topology. The relation $x \prec y$ behaves somewhat like a strict ordering relation; at least its meaning is clearly that y should be *definitely* larger than in the partial ordering. Such a relation has many pleasant properties. The primary purpose of introducing it is to provide a simple definition for the kind of spaces that are most useful to us. We first mention the most elementary features of this relation.

2.2 Proposition. *In a complete lattice* D *we have:*

(i) $\perp \prec x$;

(ii) $x \prec z$ *and* $y \prec z$ *imply* $x \sqcup y \prec z$;

(iii) $x \prec y \sqsubseteq z$ *implies* $x \prec z$;

(iv) $x \sqsubseteq y \prec z$ *implies* $x \prec z$;

(v) $x \prec y$ *implies* $x \sqsubseteq y$;

(vi) $x \prec x$ *iff* $\{z \in D : x \sqsubseteq z\}$ *is open;*

(vii) *if* $S \subseteq D$ *is directed, then*

$x \prec \bigsqcup S$ *iff* $x \prec y$ *for some* $y \in S$. □

The proofs of these statements can be safely left to the reader.

2.3 Definition. A *continuous lattice* is a complete lattice D in which for every $y \in D$ we have:

$$y = \bigsqcup\{x \in D : x \prec y\}.$$

As an alternate definition we find:

2.4 Proposition. *A complete lattice* D *is continuous iff for every* $y \in D$ *we have:*

$$y = \bigsqcup\{\textstyle\bigsqcap U : y \in U\},$$

where U *ranges over the open subsets of* D.

Proof: Suppose D is continuous. If $y \in D$ and $x \prec y$, then let

$$U = \text{Int } \{z : x \sqsubseteq z\},$$

an open set. Now $y \in U$ by definition, and

$$U \subseteq \{z : x \sqsubseteq z\}.$$

Thus,

$$x \sqsubseteq \textstyle\bigsqcap U \sqsubseteq y.$$

It easily follows by lattice theory that the equation of 2.3 implies that of 2.4.

In the converse direction we have only to note that if U is open and $y \in U$, then $\bigsqcap U \prec y$. The implication from 2.4 to 2.3 results at once. □

What is the idea of this definition? A continuous lattice is more special than a complete lattice: not only are lubs to be limits but every element must be a limit *from below*. This rather rough remark can be made more precise. In any complete lattice D define the *principal limit* of a net $x : I \to D$ by the formula:

$$lim\langle x_i : i \in I \rangle = \bigsqcup\{\bigsqcap\{x_j : j \geq i\} : i \in I\}.$$

Then specify that x *converges* to $y \in D$ iff

$$y \sqsubseteq lim\langle x_i : i \in I\rangle .$$

Having a notion of convergence, we can then say that $U \subseteq D$ is *open* iff every net converging to an element of U is eventually in U. This gives nothing more than what we have called the induced topology above, as is easily checked. But now being in possession of a topology, we can redefine convergence in the usual way. Question: when do the two notions of convergence agree? Answer: if and only if D is a continuous lattice.

For obviously by construction the limit definition of convergence implies the topological. Now if D is a continuous lattice and x converges to y topologically, consider an open $U \subseteq D$ with $y \in U$. For some $i \in I$ we shall have $x_j \in U$ for all $j \geq i$. Therefore

$$\bigsqcap U \sqsubseteq \bigsqcap\{x_j : j \geq i\} \sqsubseteq lim\langle x_i : i \in I\rangle .$$

From the formula of 2.4 it at once follows that $y \sqsubseteq lim\langle x_i : i \in I\rangle$. Thus, in continuous lattices, we have shown that the two notions of convergence are the same. Finally, suppose that the two notions coincide for a complete lattice D. Define a set $I = \{(U,z) : y,z \in U\}$, where z ranges over D and U over open subsets of D. This set is *directed* by the relation: $(U,z) \leq (V,w)$ iff $U \supseteq V$. Let $x : I \to D$ be given by: $x(U,z) = z$. Then x is a net converging to y topologically. But $lim\langle x_i : i \in I\rangle = \bigsqcup\{\bigsqcap U : y \in U\}$. In this way we see that the assumption about the two styles of convergence implies that D is a continuous lattice in view of 2.4.

In T_0-spaces continuous functions are always *monotonic* (i.e. $\sqsubseteq$-preserving). For continuous lattices, by virtue of the remarks we have just made about limits, we can define the continuity of $f : D \to D'$ to mean that $f(lim\langle x_i : i \in I\rangle) \sqsubseteq lim'\langle f(x_i) : i \in I\rangle$ for all nets $x : I \to D$. This is all very fine, but general limits are messy to work with; we shall find it easier to state results in terms of lubs as in 2.5-2.7 below.

Before going any deeper, however, we should clear up another point about topologies. Suppose that D is any T_0-space which becomes a complete lattice under its induced partial ordering. Then it is evident from our definitions that every set open in the given topology is also open in the topology induced from the lattice structure. Question: when do the two topologies agree? Answer: a *sufficient* condition is that the equation:

$$y = \bigsqcup\{\bigsqcap U : y \in U\}$$

hold for all $y \in$ D, where U ranges over the *given* open sets. Because in that case if V is open in the lattice sense and $y \in V$, then $\sqcap U \in V$ for some set U, open in the given sense, where $y \in U$. But $U \subseteq V$ follows, and so V is a union of given open sets and is itself open in the given topology. Of course this equation implies that D is a continuous lattice by virtue of 2.4. Notice that by the same token the sets of the form $\{y \in \mathrm{D} : x \prec y\}$ will form a basis for the open sets of a continuous lattice.

2.5 Proposition. *If* D *and* D′ *are complete lattices with their induced topologies, then a function* $f:\mathrm{D} \to \mathrm{D}'$ *is continuous iff for all directed subsets* $S \subseteq \mathrm{D}$:

$$f(\bigsqcup S) = \bigsqcup\{f(x) : x \in S\}.$$

Proof: If $f:\mathrm{D} \to \mathrm{D}'$ is continuous, the equation follows from the definition of continuous function and the fact that lubs are limits. Assume then that the equation holds for all directed sets S. Let $U' \subseteq \mathrm{D}'$ be open in D′ and let

$$U = \{x \in \mathrm{D} : f(x) \in U'\}.$$

We must show that U is open in D. Note first that if $x \sqsubseteq y$, then

$$S = \{x,y\}$$

is directed; hence,

$$f(x \sqcup y) = f(y) = f(x) \sqcup f(y),$$

so $f(x) \sqsubseteq f(y)$. Thus f is monotonic and so U satisfies condition (i). That U satisfies condition (ii) follows at once from the above equation. □

2.6 Proposition. *With functions from complete lattices to complete lattices, a function of several variables is continuous in the variables jointly iff it is continuous in the variables separately.*

Proof: It will be sufficient to discuss functions of two variables. The product D×D′ of two complete lattices is a complete lattice, and it is easy to check that the induced topology is the product topology. Since projection is continuous, joint continuity implies separate continuity. To check the converse suppose that

$$f:\mathrm{D}\times\mathrm{D}' \to \mathrm{D}''$$

is a map where the separate continuity holds as follows:

$$f(\bigsqcup S,y) = \bigsqcup\{f(x,y) : x \in S\}$$

and

$$f(x,\bigsqcup S') = \bigsqcup\{f(x,y) : y \in S'\}$$

where $S \subseteq D$ and $S' \subseteq D'$are directed and $x \in D$ and $y \in D'$. Let now

$$S^* \subseteq D \times D'$$

be directed in the product. The projection of S^* to $S \subseteq D$ and S^* to $S' \subseteq D'$ produces directed subsets of D and D'.
Note that

$$\bigsqcup S^* = (\bigsqcup S, \bigsqcup S').$$

Thus by assumption

$$f(\bigsqcup S^*) = \bigsqcup\{f(x,y) : x \in S,\ y \in S'\}.$$

But since S^* is directed, $x \in S$ and $y \in S'$ implies $x \sqsubseteq u$ and $y \sqsubseteq v$ for $(u,v) \in S^*$. Thus by monotonicity of f we can show

$$f(\bigsqcup S^*) = \bigsqcup\{f(u,v) : (u,v) \in S^*\}$$

and that gives the joint continuity. □

One of the justifications (by euphony at least) of the term *continuous lattice* is the fact that such spaces allow for *so many* continuous functions. One indication of this is the result:

2.7 Proposition. *In a continuous lattice* D *the finitary lattice operations* $\sqcup$ *and* $\sqcap$ *are continuous.*

Proof: It is trivial to show that $\sqcup$ is continuous in every complete lattice; this is not so for $\sqcap$. In view of 2.6 we need only show

$$x \sqcap \bigsqcup S = \bigsqcup\{x \sqcap y : y \in S\}$$

for every directed $S \subseteq D$. In fact it is enough to show

$$x \sqcap \bigsqcup S \sqsubseteq \bigsqcup\{x \sqcap y : y \in S\}$$

because the opposite inequality is valid in all complete lattices. In view of the fact that D is *continuous*, it is enough to show that

$$t \prec x \sqcap \bigsqcup S \text{ implies } t \sqsubseteq \bigsqcup\{x \sqcap y : y \in S\}.$$

So assume $t \prec x \sqcap \bigsqcup S$. Then $t \sqsubseteq x \sqcap \bigsqcup S \sqsubseteq x$. Also $t \prec \bigsqcup S$ because $x \sqcap \bigsqcup S \sqsubseteq \bigsqcup S$. Thus $t \prec y$ for some $y \in S$ since the set

$$\{z \in D : t \prec z\}$$

is *open*. But then $t \sqsubseteq y$, and so $t \sqsubseteq x \sqcap y$, and the result follows. □

It is now time to provide some examples of continuous lattices.

<u>2.8 Proposition</u>. *A finite lattice is a continuous lattice.* □

<u>2.9 Proposition</u>. *The Cartesian product of any number of continuous lattices is a continuous lattice with the induced topology agreeing with the product topology.*

<u>2.10 Proposition</u>. *A retract of a continuous lattice is a continuous lattice with the subspace topology agreeing with the induced topology.*

It would seem that the continuous lattices are starting to sound suspiciously like the injective spaces. Indeed, if we can prove the following, the circle will be complete.

<u>2.11 Proposition</u>. *Every continuous lattice is an injective space under its induced topology.*

<u>2.12 Theorem</u>. *The injective spaces are exactly the continuous lattices.*

This theorem is an immediate consequence of the preceding results: an injective space is a retract of a power of $\mathbb{O}$. But $\mathbb{O}$ is a finite lattice ($\bot \sqsubseteq \top$), and so the given space is a continuous lattice under its induced topology. On the other hand a continuous lattice is injective. It remains then to prove 2.9 - 2.11.

Proof of 2.9 Let D_i for $i \in I$ be a system of continuous lattices. The product

$$D^* = \bigtimes_{i \in I} D_i$$

is a complete lattice in the usual way and has its induced topology. Suppose $y \in D^*$ and let $i \in I$. Then $y_i \in D_i$. Since D_i is a continuous lattice

$$y_i = \bigsqcup \{x \in D_i : x \prec y_i\}.$$

For $x \in D_i$, let $[x]^i \in D^*$ be defined by:

$$[x]^i_j = \begin{cases} x & \text{if } i=j, \\ \bot & \text{if } i \neq j. \end{cases}$$

Note that since D_i is continuous we have:

$$[y_i]^i = \bigsqcup \{[x]^i : x \prec y_i\},$$

and

$$y = \bigsqcup \{[y_i]^i : i \in I\}.$$

It follows that

$$y = \bigsqcup\{\bigsqcap\{z : z_i \in U\} : i \in I,\ y_i \in U\},$$

where i ranges over I and U over the open subsets of D_i, because

$$[x]^i \sqsubseteq \bigsqcap\{z : z_i \in U\}, \text{ where } U = \{u \in D_i : x \prec u\}.$$

But the sets $\{z : z_i \in U\}$ are open in the product sense, and so

$$y = \bigsqcup\{\bigsqcap U : y \in U\},$$

where U ranges now over the open subsets of the product topology on D^*. By the remark following 2.4 we conclude that D^* is continuous with the lattice-induced topology being the product topology. □

Proof of 2.10 Let D′ be a continuous lattice and let D ⊆ D′ be a subspace which is a retract. We have for a suitable $j : D' \to D$,

$$D = \{x \in D' : j(x) = x\},$$

where of course j is continuous.

First a note of *warning:* though D is a subspace it is *not* a sublattice; that is, the partial ordering on D is the restriction of that of D′, but the lubs of D are not those of D′. We shall have to distinguish operations by adding a prime (′) for those of D′.

Suppose $x, y \in D$. Let $z' = x \sqcup' y \in D'$ and define $z = j(z') \in D$. Now $x \sqsubseteq z'$ and $y \sqsubseteq z'$ and j is monotonic, so $x \sqsubseteq z$ and $y \sqsubseteq z$. Suppose $x \sqsubseteq w$ and $y \sqsubseteq w$ with $w \in D$. Then in D′ we have $x \sqcup' y \sqsubseteq w$; so $z \sqsubseteq w$ also. Hence we have shown that $z = x \sqcup y$ in D.

To show that D has a least element $\bot$ (which may be larger than the $\bot' \in D'$), we need a well-known lemma about monotonic functions: Every monotonic function on a complete lattice into itself has a *least* fixed point. (Cf. Birkhoff (1970), p. 115.) In our case j is monotonic and

$$\bot = \bigsqcap{}'\{x \in D' : j(x) \sqsubseteq x\}$$

is the desired element in D.

Thus D is at least a semilattice with $\bot$ and $\sqcup$. To show that D is a lattice we need to show that every directed $S \subseteq D$ has a lub in D. Now we know:

$$\bigsqcup{}' S \in D',$$

and this is a *limit* of a monotone net. So by 2.1, and the continuity of j:

$$j(\bigsqcup{}' S) = \bigsqcup\{j(x) : x \in S\} = \bigsqcup S$$

in D. In this way we now know that D is a complete lattice. We must still show that D is continuous.

Suppose $y \in D$. In D' we can write:

$$y = \bigsqcup{}'\{x \in D' : x \prec y\}$$

and this is the limit of a monotone net. Thus

$$j(y) = y = \bigsqcup\{j(x) : x \prec y,\ x \in D'\},$$

where the lub is taken in D. Note that the sets

$$U = \{z \in D : x \prec z\}$$

are open in D for each $x \in D'$. Note too that if $z \in U$, then $x \sqsubseteq z$ and so $j(x) \sqsubseteq j(z) = z$. This means that

$$j(x) \sqsubseteq \bigsqcap U$$

in D. We can then write in D:

$$y = \bigsqcup\{\bigsqcap U : y \in U\}$$

where U ranges over the open subsets of D, and so the lattice is continuous by 2.4. Inasmuch as the open sets U just used were open in the subspace topology, it follows by the remark after 2.4 that the subspace and the lattice-induced topologies coincide. □

Proof of 2.11: Let D be a continuous lattice with its induced topology, and let $X \subseteq Y$ be two T_0-spaces in the subspace relation. Suppose

$$f : X \to D$$

is continuous. Define

$$\bar{f} : Y \to D$$

by the formula:

$$\bar{f}(y) = \bigsqcup\{\bigsqcap\{f(x) : x \in X \cap U\} : y \in U\},$$

where U ranges over the open subsets of Y. We need to show that $\bar{f}$ extends f and that it is continuous.

First, the continuity: Suppose that $d \in D$ and $d \prec \bar{f}(y)$; that is, $\bar{f}(y)$ belongs to a typical basic open subset of D. Since D is continuous, we can also find

$$d \prec d' \prec \bar{f}(y)$$

with $d' \in D$. From the definition of $\bar{f}$ it follows that

$$d' \sqsubseteq \bigsqcap\{f(x) : x \in X \cap U\}.$$

for some open $U \subseteq Y$ with $y \in U$. Thus

$$d' \sqsubseteq \bar{f}(y')$$

for *all* $y' \in U$ by virtue of the definition of $\bar{f}$. Since $d \prec d'$, we have also

$$d \prec \bar{f}(y')$$

for all $y' \in U$; in other words, the inverse image of the open subset of D determined by d under $\bar{f}$ is indeed open in Y, and $\bar{f}$ is thus continuous.

Next, the extension property: Note that the relationship

$$\bar{f}(x') \sqsubseteq f(x')$$

for all $x' \in X$ comes directly out of the definition of $\bar{f}$. For the converse, suppose $d \prec f(x')$ where $d \in D$. By assumption $f : X \to D$ is continuous, so

$$d \prec f(x'')$$

for all $x'' \in X \cap U$ where U is a suitable open subset of Y with $x' \in U$. In particular we have:

$$d \sqsubseteq \bigsqcap\{f(x'') : x'' \in X \cap U\},$$

and so $d \sqsubseteq \bar{f}(x')$. Since $d \prec f(x')$ always implies $d \sqsubseteq \bar{f}(x')$, we see that $f(x') \sqsubseteq \bar{f}(x')$ follows by the continuity of D, and the proof is complete. □

The lattice approach to injective spaces gives a completely "internal" characterization of them: in the first place the lattices are complete. Next we can define lattice theoretically:

$x \prec y$ iff whenever $y \sqsubseteq \bigsqcup Z$ and $Z \subseteq D$ is directed, then $x \sqsubseteq z$ for some $z \in Z$.

Finally we assume that for all $y \in D$:

$$y = \bigsqcup\{x \in D : x \prec y\}.$$

That makes D a continuous lattice with the sets $\{y \in D : x \prec y\}$ as a basis for the topology. Such T_0-spaces are injective and every injective space can be obtained in this way with the lattice structure being uniquely determined by the topology. Furthermore, as we have seen, the injective property can be exhibited, as in the proof of 2.11, by an explicit formula for extending functions.

The retract approach to injective spaces should also be considered. The Cartesian powers $\mathbb{O}^I$ are very simple spaces; indeed, as lattices these are just the Boolean algebras of *all subsets of* I (that is, isomorphic thereto). The topology has as a basis the classes of sets containing given finite sets (the *weak* topology, cf. Nerode (1959)). A continuous function

$$j : \mathrm{P}I \to \mathrm{P}I$$

is one of "finite character" so that

$$j(X) = \bigcup\{j(F) : F \subseteq X\}$$

where $X \subseteq I$ and F ranges over *finite* sets. Such a function j is a *retraction* iff it is an idempotent:

$$j \circ j = j,$$

which means that the range of j *is* the set of fixed points of j. As we have seen

$$D = \{X \in \mathrm{P}I : j(X) = X\}$$

is a continuous lattice (under $\subseteq$ in this case), and *every* continuous lattice is isomorphic to one obtained in this way. This provides a representation theorem of sorts for continuous lattices, but it does not seem to be of too much help in proving theorems.

The reader should not forget that any space (any *given* number of spaces X, Y,...) can be found as a *subspace* of a continuous lattice D. Since D is injective any continuous function $f : X \to Y$ can be extended to a continuous function $\bar{f} : D \to D$. Thus the continuous functions from D into D are a rich totality including *all* the structure of continuous functions on *all* the subspaces. And this remark brings us to the study of function spaces.

3. FUNCTION SPACES. We recall the standard definition and introduce our notation for function spaces.

3.1 Definition. For T_0-spaces X and Y we let $[X \to Y]$ be the *space of all continuous functions* f: $X \to Y$ endowed with the product topology, sometimes called the topology of pointwise convergence. This topology has as a subbase sets of the form:

$$\{f : f(x) \in U\}$$

where $x \in X$ and $U \subseteq Y$ is open.

The pointwise aspect of the topology is immediately apparent in the partial ordering.

3.2 Proposition. *The induced partial ordering on* $[X \to Y]$ *is such that:*

$$f \sqsubseteq g \quad \textit{iff} \quad f(x) \sqsubseteq g(x) \textit{ for all } x \in X,$$

where $f, g \in [X \to Y]$. □

The first question, of course, is what kind of a partial ordering this is.

<u>3.3 Theorem</u>. *If* D *and* D′ *are continuous lattices, then so is* $[D \to D']$ *under the induced partial ordering with the lattice topology agreeing with the product topology.*

Proof: The argument is "pointwise." Thus, the constant function with value $\bot \in D'$ is obviously continuous and is the $\bot$ of $[D \to D']$ by 3.2. Since by 2.7 the lattice operation $\sqcup$ on D′ is continuous, then if $f, g \in [D \to D']$ the composition $f \sqcup g$, defined by

$$(f \sqcup g)(x) = f(x) \sqcup g(x)$$

for all $x \in D$, is also continuous and represents the lub of $\{f, g\}$ in $[D \to D']$. (The same arguments apply to $\top$ and $\sqcap$, so $[D \to D']$ is at least a lattice.) To show that $[D \to D']$ is complete it is sufficient now to show that lubs of *directed* subsets exist. So let $\mathcal{F} \subseteq [D \to D']$ be directed. Define a function from D into D′ by the equation:

$$(\bigsqcup \mathcal{F})(x) = \bigsqcup \{f(x) : f \in \mathcal{F}\},$$

for all $x \in D$. If we can show that $\bigsqcup \mathcal{F}$ is continuous, then being in $[D \to D']$ it has to be the lub. Consider $U \subseteq D'$, an open subset. Taking the inverse image and remembering that $\mathcal{F}$ is directed, we find:

$$\{x : (\bigsqcup \mathcal{F})(x) \in U\} = \bigcup \{\{x : f(x) \in U\} : f \in \mathcal{F}\}.$$

This is an open set, and so $\bigsqcup \mathcal{F}$ is indeed continuous. (*Warning:* the infinite $\bigsqcap \mathcal{F}$ are *not* in general computed pointwise; however, it is easy to extend the above argument to show that arbitrary $\bigsqcup \mathcal{F}$ are.)

To show that $[D \to D']$ is continuous, we establish first that for $f \in [D \to D']$

$$f = \bigsqcup \{\vec{e}[e, e'] : e' \prec f(e)\},$$

where e ranges over D and e' over D′, and where the function $\vec{e}[e, e']$ is defined by:

$$\vec{e}[e, e'](x) = \begin{cases} e' & \text{if } e \prec x, \\ \bot & \text{if not,} \end{cases}$$

for all $x \in D$. Call the function on the right f'. Calculate:

$$f'(x) = \bigsqcup\{\vec{e}[e,e'](x) : e' \prec f(e)\}$$
$$= \bigsqcup\{e' : \exists e \prec x[e' \prec f(e)]\}$$
$$= \bigsqcup\{e' : e' \prec f(x)\} = f(x).$$

With the equation for f proved, note next that for all $g \in [D \to D']$,

$$e' \prec g(e) \text{ implies } \vec{e}[e,e'] \sqsubseteq g$$

by an easy pointwise argument. If we let

$$V = \{g : e' \prec g(e)\},$$

we see then that V is open in the product topology and that

$$\vec{e}[e,e'] \sqsubseteq \bigsqcap V.$$

We may then conclude that

$$f = \bigsqcup\{\bigsqcap V : f \in V\},$$

which proves both that $[D \to D']$ is a continuous lattice and that the two topologies agree by the remark following 2.4. □

The above theorem might possibly be generalized to $[X \to D]$ where X is merely a T_0-space, but I was unable to see the argument. In any case we are mostly interested in the continuous lattices. Note as a consequence of our proof:

3.4 Corollary. *For continuous lattices* D *and* D', *the evaluation map:*

$$\text{eval} : [D \to D'] \times D \to D'$$

is continuous.

Proof: Here $\text{eval}(f,x) = f(x)$. With f fixed, this is obviously continuous. With x fixed, we proved the continuity above with our calculation of $\bigsqcup \mathcal{F}$ in view of 2.5. Hence applying 5.3 and 2.6, we conclude that eval is jointly continuous. □

This result gives only one example of the masses of continuous functions that are available on continuous lattices. As another fundamental example we have:

3.5 Proposition. *For continuous lattices* D, D', *and* D", *the map of functional abstraction:*

$$\text{lambda} : [[D \times D'] \to D''] \to [D \to [D' \to D'']]$$

is continuous.

Proof: Here lambda is defined by:

$$\mathrm{lambda}(f)(x)(y) = f(x,y)$$

where $f \in [[D \times D'] \to D'']$ and $x \in D$ and $y \in D'$. What is particularly interesting here is that by virtue of 3.3 we are making use of $[D \to [D' \to D'']]$ as a continuous lattice. The principle being stated here can be formulated more broadly in this way:

> If an expression $\mathcal{E}(x,y,z,\ldots)$ is continuous in all its variables $x,y,z,\ldots$ with values in D' as x ranges in D, then the expression
>
> $$\lambda x{:}D.\mathcal{E}(x,y,z,\ldots)$$
>
> with values in $[D \to D']$ is continuous in the remaining variables $y,z,\ldots$.

The λ-notation is a notation for *functions*, where in the above the variable after the λ is the *argument* and the expression after the . is the *value* (as a function of the argument). Thus we could write:

$$\mathrm{lambda} = \lambda f{:}[[D \times D'] \to D''].\lambda x{:}D.\lambda y{:}D'.f(x,y),$$

and, because $f(x,y)$ is continuous in f, x, and y, our conclusion follows. But often it is more readable not to write equations between functions but rather equations between values for definitional purposes.

The proof of the principle is easy. For let the variable y, say range over D'' and let $S \subseteq D''$ be a directed subset. Then

$$\lambda x{:}D.\mathcal{E}(x, \bigsqcup S, z,\ldots) = \lambda x{:}D.\bigsqcup\{\mathcal{E}(x,y,z,\ldots) : y \in S\}$$
$$= \bigsqcup\{\lambda x{:}D.\mathcal{E}(x,y,z,\ldots) : y \in S\},$$

because the lubs of functions are computed pointwise. □

We need not enumerate the many corollaries that follow easily now from this result. We mention, however, that composition $f\circ g$ of functions (on continuous lattices) is continuous in the two function variables, where we write

$$(f\circ g)(x) = f(g(x)).$$

What will be useful will be to return at this point to a discussion of the injective properties of continuous lattices. If one continuous lattice is a subspace of another it is of course a retract. This relationship between spaces can be given by a pair of continuous maps

$$i : D \to D' \quad \text{and} \quad j : D' \to D ,$$

where

$$j \circ i = \mathrm{id}_D = \lambda x{:}D.x \ .$$

The composition $i \circ j : D' \to i(D)$ is the retraction onto the subspace corresponding to D under i. Now if we have a diagram:

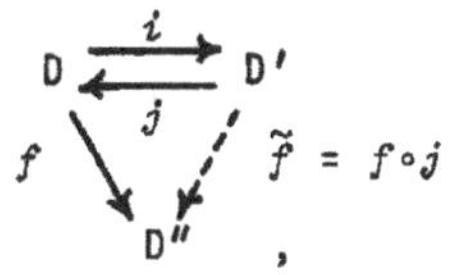

the given continuous f is at once extendable from D to D′ by the obvious definition of $\tilde{f}$. This $\tilde{f}$ is *not* the $\bar{f}$ used in the proof of 2.11, and it will be well to sort out the connections. On one side note that if f' is any function which extends f, then we have $f = f' \circ i$. But this implies

$$\tilde{f} = f \circ j = f' \circ i \circ j ,$$

which shows that $\tilde{f}$ is a "degraded" version of f'. There is one situation where this type of degrading is especially nice.

3.6 Definition. A continuous lattice D is said to be a *projection* of a continuous lattice D′ iff there are a pair of continuous maps

$$i : D \to D' \text{ and } j : D' \to D$$

such that not only

$$j \circ i = \mathrm{id}_D ,$$

but also

$$i \circ j \sqsubseteq \mathrm{id}_{D'} .$$

Thus, in case our retraction is a projection, we have $\tilde{f} \sqsubseteq f'$, which means that $\tilde{f}$ is the *minimal* extension of $f \in [D \to D'']$ to a function in $[D' \to D'']$. We will discuss the nature of $\tilde{f}$ in a moment. But before we do we pause to remark that the correspondence $f \leadsto \tilde{f}$ is *continuous*, and this fact is easily extended.

3.7 Proposition. *Suppose the two pairs of maps*

$$i_n : D_n \to D'_n \text{ and } j_n : D'_n \to D_n$$

for $n = 0,1$ *make* D_n *a retraction (projection) of* D'_n. *Then* $[D_0 \to D_1]$ *is also a retraction (projection) of* $[D'_0 \to D'_1]$ *by means of the pair of maps:*

$$\vec{i}(f) = i_1 \circ f \circ j_0 \text{ , and}$$

$$\vec{j}(f') = j_1 \circ f' \circ i_0 \text{ ,}$$

where $f \in [D_0 \to D_1]$ *and* $f' \in [D'_0 \to D'_1]$. □

Returning now to $\bar{f}$ we can prove:

<u>3.8 Proposition</u>. *If* D *is a continuous lattice and* $e : X \to Y$ *a subspace embedding, then for each* $f : X \to D$, *the function* $\bar{f} : Y \to D$ *given by the formula*

$$\bar{f}(y) = \bigsqcup\{\bigsqcap\{f(x) : e(x) \in U\} : y \in U\},$$

where U *ranges over open subsets of* Y *and* x *over* X, *is the maximal extension of* f *to a function in the partially ordered set* $[Y \to D]$.

<u>*Proof:*</u> We are saying that $\bar{f}$ is the maximal solution to the equation

$$f = \bar{f} \circ e \text{ .}$$

We already know it is a solution, so let f' be any other. We have

$$\begin{aligned} f'(y) &= \bigsqcup\{\bigsqcap\{f'(z) : z \in U\} : y \in U\} \\ &\sqsubseteq \bigsqcup\{\bigsqcap\{f'(z) : z \in e(X) \cap U\} : y \in U\} \\ &= \bigsqcup\{\bigsqcap\{f'(e(x)) : e(x) \in U\} : y \in U\} \\ &= \bigsqcup\{\bigsqcap\{f(x) : e(x) \in U\} : y \in U\} \\ &= \bar{f}(y) \text{ ,} \end{aligned}$$

which establishes that $f' \sqsubseteq \bar{f}$. □

By the same argument we could show that $\bar{f}$ is the maximal solution of $\bar{f} \circ e \sqsubseteq f$. An interesting question is whether the correspondence $f \rightsquigarrow \bar{f}$ is *continuous*. I very much doubt it, but at this moment a counterexample escapes me. It is clear that the correspondence is *monotonic*, for if $f \sqsubseteq g$, then the formula of 3.8 shows that $\bar{f} \sqsubseteq \bar{g}$. This gives us a neat argument for the previous remark: if $g \circ e \sqsubseteq f$, then

$$\overline{g \circ e} \sqsubseteq \bar{f}$$

But $g \circ e = g \circ e$, so by 3.8, $g \sqsubseteq \overline{g \circ e}$, and $\bar{f}$ is thus maximal.

In the case that the range spaces are being extended, the following lemma relating the extensions will be very useful when we consider inverse limits.

3.9 Lemma. *Consider the diagram:*

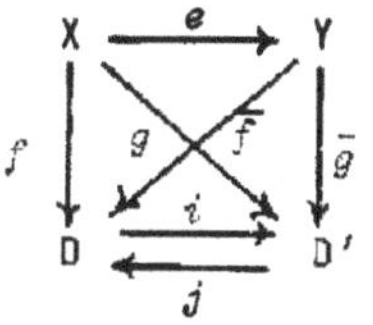

where the upper row is a subspace embedding and the lower is a projection. If the given functions f and g are extended to $\bar{f}$ *and* $\bar{g}$ *as in* 3.8, *and if* $f = j \circ g$, *then* $\bar{f} = j \circ \bar{g}$ *also.*

Proof: $\bar{f}$ and $\bar{g}$ are maximal solutions of $f = \bar{f} \circ e$ and $g = \bar{g} \circ e$. Therefore since

$$f = j \circ g = j \circ \bar{g} \circ e,$$

we see that

$$j \circ \bar{g} \sqsubseteq \bar{f}.$$

Note also that

$$i \circ \bar{f} \circ e = i \circ f = i \circ j \circ g \sqsubseteq g,$$

and so by the remark following 3.8, we have

$$i \circ \bar{f} \sqsubseteq \bar{g}.$$

Therefore

$$\bar{f} = j \circ i \circ \bar{f} \sqsubseteq j \circ \bar{g},$$

which proves the equality. □

Whether this lemma is true for retractions in any form, I do not know. My proof seems to require the stronger projection relationship. I suspect there may be difficulties. In general projections are better behaved than retractions. By the way the word *projection* seems to be properly used in 3.6, for the projection $j: D \times D' \to D$ of the Cartesian product of two continuous lattices onto the first factor *is* a projection with partial inverse $i: D \to D \times D'$ defined by

$$i(x) = (x, \perp)$$

for $x \in D$.

3.10 Proposition. *If the continuous lattice* D *is a projection of the continuous lattice* D′ *via the pair of maps* i, j; *then for all* $S \subseteq D$ *and all* $x, y \in D$ *we have:*

(i) $i(\bigsqcup S) = \bigsqcup\{i(x) : x \in S\}$,

(ii) $i(x) = i(y)$ *implies* $x=y$,

(iii) $x \prec y$ *implies* $i(x) \prec i(y)$.

Conversely, if a map $i: D \to D'$ *satisfies* (i) - (iii), *then there exists a continuous* $j: D' \to D$ *making* D *a projection of* D′, *and in fact* j *is uniquely determined by:*

(iv) $j(x') = \bigsqcup\{x \in D : i(x) \sqsubseteq x'\}$

for all $x' \in D'$.

Proof: Equation (i) holds for *directed* $S \subseteq D$ because i is continuous. To have it hold for arbitrary S it is only necessary to check it for *finite* sets, because every lub *is* the directed lub of finite sublubs. (The last word of that sentence is an unfortunate accident.) Further, to check the equation for finite sets it is enough to check it for the empty set and for two element sets. Thus, $i(\bot) = \bot$, because $j(i(\bot)) = \bot$ and since $\bot \sqsubseteq i(\bot)$,

$$j(\bot) \sqsubseteq j(i(\bot)) = \bot,$$

so $j(\bot) = \bot$. Whence $i(\bot) = i(j(\bot)) \sqsubseteq \bot$. Next $i(x \sqcup y) = i(x) \sqcup i(y)$, because first

$$i(x) \sqcup i(y) \sqsubseteq i(x \sqcup y)$$

by monotonicity. Then note that

$$i(x) \sqsubseteq i(x) \sqcup i(y)$$

and so

$$x = j(i(x)) \sqsubseteq j(i(x) \sqcup i(y)).$$

Similarly

$$y \sqsubseteq j(i(x) \sqcup i(y)),$$

whence

$$x \sqcup y \sqsubseteq j(i(x) \sqcup i(y)).$$

But then

$$i(x \sqcup y) \sqsubseteq i(j(i(x) \sqcup i(y))) \sqsubseteq i(x) \sqcup i(y),$$

which completes the argument for equation (i).

Condition (ii) is obvious. For (iii) we argue as follows. Assume $x \prec y$. Since D′ is continuous we can write:

$$i(y) = \bigsqcup \{z' \in D' : z' \prec i(y)\},$$

and conclude by the continuity of j that:

$$y = j(i(y)) = \bigsqcup \{j(z') : z' \prec i(y)\}.$$

But $x \prec y$, so $x \prec j(z')$ for some $z' \prec i(y)$. Now $x \sqsubseteq j(z')$ follows; therefore $i(x) \sqsubseteq i(j(z')) \sqsubseteq z'$. Thus $i(x) \prec i(y)$.

Turning now to the converse, assume of the map i that it s satisfies (i) - (iii). Compute:

$$i(j(x')) = \bigsqcup \{i(x) : i(x) \sqsubseteq x'\} \sqsubseteq x'.$$

This is correct because i is continuous and the set $\{x : i(x) \sqsubseteq x'\}$ is directed in view of condition (i). Thus $i \circ j \sqsubseteq \mathrm{id}_{D'}$. Note that by virtue of (i) and (ii) it is the case that

$$i(x) \sqsubseteq i(y) \text{ implies } x \sqsubseteq y.$$

(The reason is that $x \sqsubseteq y$ is equivalent to $x \sqcup y = y$.) This remark allows us to compute:

$$j(i(y)) = \bigsqcup \{x : i(x) \sqsubseteq i(y)\}$$
$$= \bigsqcup \{x : x \sqsubseteq y\} = y.$$

Hence, $j \circ i = \mathrm{id}_D$. It remains to show that j is continuous.

Suppose $S' \subseteq D'$ is directed. Since j is by definition monotonic, it is sufficient to prove that

$$j(\bigsqcup S') \sqsubseteq \bigsqcup \{j(x') : x' \in S'\}.$$

Now

$$j(\bigsqcup S') = \bigsqcup \{x : i(x) \sqsubseteq \bigsqcup S'\},$$

so suppose $i(x) \sqsubseteq \bigsqcup S'$. Let $z \prec x$; whence $i(z) \prec i(x)$. Thus $i(z) \prec x'$ for some $x' \in S$, and therefore $i(z) \sqsubseteq x'$. We obtain then $z \sqsubseteq j(x')$, which means that

$$z \sqsubseteq \bigsqcup \{j(x') : x' \in S'\}$$

holds for all $z \prec x$. By the continuity of D we conclude

$$x \sqsubseteq \bigsqcup \{j(x') : x' \in S'\}$$

holds for all x with $i(x) \sqsubseteq \bigsqcup S'$. The desired result follows. □

As a corollary of 3.10 we can easily see which *subspaces* of a continuous lattice D′ are projections of it. Such a subspace $D \subseteq D'$ must first be closed under $\bigsqcup$. That is, if $S \subseteq D$, then $\bigsqcup S \in D$ for all S, where the lub is taken in the sense of D'. The identity map on D will then satisfy (i) and (ii). But this is not enough, since we would not know that D is a continuous lattice, nor whether (iii) holds. The following additional condition would be sufficient, if assumed for all $y \in$ D:

$$y = \bigsqcup \{x \in D : x \prec y\},$$

where $\prec$ is taken in the sense of D'. This implies that

$$y = \bigsqcup \{\bigsqcap (D \cap U) : y \in U\}$$

where U ranges over the open subsets of D' and where the $\bigsqcap$ is taken in the sense of D. This condition makes the subspace topology the same as the lattice topology on D and besides makes D continuous, which is just what we need. (Another way to put it is that whenever $z \prec y$, where $y \in$ D but $z \in$ D', then $z \sqsubseteq x \prec y$, for some $x \in$ D.)

It seems a bit troublesome to characterize in a simple way just which maps j:D'$\to$ D are projections. (Other than saying outright that the map i:D $\to$ D' such that for all $x \in$ D:

$$i(x) = \bigsqcap \{x' \in D' : x \sqsubseteq j(x')\}$$

is the continuous partial inverse to j.) But we can say very easily which continuous maps j:D'$\to$ D' are projections onto *subspaces*; namely, we must have

$$j = j \circ j \sqsubseteq \mathrm{id}.$$

The subspace in question then is:

$$D = \{x \in D' : j(x) = x\}.$$

This non-empty subspace is the exact range of j and is closed under $\bigsqcup$. Let $y \in$ D. Then if $x' \prec y$ in D', we find $j(x') \sqsubseteq x' \prec y$. Thus since

$$y = \bigsqcup \{x' \in D' : x' \prec y\},$$

we see that

$$y = j(y) = \bigsqcup \{j(x') : x' \prec y\}.$$

But each $j(x) \in$ D, so $y = \bigsqcup \{x \in D : x \prec y\}$, as desired.

The foregoing discussion suggests looking more closely at the space of all projections of a continuous lattice since they are so easily characterized.

<u>3.11 Definition</u>. Given a continuous lattice D, we let the *space of projections* be denoted by:

$$J_D = \{j \in [D \to D] : j = j \circ j \sqsubseteq \mathrm{id}\}.$$

<u>3.12 Proposition</u>. *For a continuous lattice* D *the space* J_D *of projections forms a complete lattice as a* $\bigsqcup$*-closed subspace of* $[D \to D]$.

<u>*Proof:*</u> The constant function $\bot \in J_D$ obviously, so J_D contains $\bigsqcup \emptyset$. Suppose $j,k \in J_D$. We wish to show that $j \sqcup k \in J_D$. Compute:

$$(j \sqcup k)((j \sqcup k)(x)) = j(j(x) \sqcup k(x)) \sqcup k(j(x) \sqcup k(x))$$

But note:

$$j(x) \sqsubseteq j(j(x)) \sqsubseteq j(j(x) \sqcup k(x)) \sqsubseteq j(x),$$

because $j(x) \sqcup k(x) \sqsubseteq x$. Similarly for $k(x)$. Therefore, we find that $(j \sqcup k) \circ (j \sqcup k) = j \sqcup k \sqsubseteq \mathrm{id}$. Suppose finally that $S \subseteq J_D$ is directed. We wish to show that $\bigsqcup S \in J_D$. Clearly $\bigsqcup S \sqsubseteq \mathrm{id}$, so compute by continuity of $\circ$:

$$\bigsqcup S \circ \bigsqcup S = \bigsqcup \{j \circ j : j \in S\} = \bigsqcup \{j : j \in S\} = \bigsqcup S.$$

It follows that J_D is $\bigsqcup$-closed and hence is a complete lattice. □

The significance of the above result becomes clearer if we consider the connection between projections and subspaces. Let us write:

$$D(j) = \{x \in D : j(x) = x\}.$$

For $j \in J_D$, each $D(j)$ is a projection of D onto a subspace. We show first that

$$j \sqsubseteq k \quad \text{iff} \quad D(j) \subseteq D(k)$$

Because if $j \sqsubseteq k$, then $j \sqsubseteq j \circ j \sqsubseteq k \circ j \sqsubseteq \mathrm{id} \circ j = j$. Therefore if $j(x) = x$, then $k(x) = k(j(x)) = j(x) = x$, which means that $D(j) \subseteq D(k)$. On the other hand, if $D(j) \subseteq D(k)$, then since $j(D) \subseteq D(j)$, we see that $k \circ j = j$. and so $j \sqsubseteq k \circ \mathrm{id} = k$. Hence J_D is isomorphic to the partially ordered set of subspaces of D that are projections. We thus conclude that these subspaces form a lattice. In fact, it is easy to show that

$$D(j \sqcup k) = \{x \sqcup y : x \in D(j), y \in D(k)\}.$$

Similarly, if S is a directed set of J_D, then $D(\bigsqcup S)$ is the $\bigsqcup$-closure in D of the subset:

$$\bigcup \{D(j) : j \in S\}.$$

These are not very deep facts, but their proofs were very much facilitated by the introduction of J_D and the utilization of the lattice structure of $[D \to D]$. Along the same line we can define for $j,k \in J_D$ a function $(j \to k) \in [D \to D] \to [D \to D]$ by the equation

$$(j \to k)(f) = k \circ f \circ j.$$

It is very easy to show that $(j \to k) \in J_{[D \to D]}$, that $(j \to k)$ is continuous in j and k, and that $[D \to D](j \to k)$ is isomorphic to $[D(j) \to D(k)]$. There are many other interesting operations on projections corresponding to other constructs besides these. And, just as with $(j \to k)$, the operations are continuous. This makes it possible to prove existence theorems about subspaces by using results like the fixed-point theorem for continuous functions. It would be even nicer if J_D turns out to be a continuous lattice itself, but as far as I can tell this is not likely to be the case.

Before we turn to the iterated function-space construction by inverse limits, there are a couple of other connections between spaces and function spaces that are useful to know.

3.13 Proposition. *Every continuous lattice* D *is a projection of its function space* $[D \to D]$.

Proof: Consider the following pair of mappings con $: D \to [D \to D]$ and min $: [D \to D] \to D$ where

$$\mathrm{con}(x)(y) = x$$

and

$$\mathrm{min}(f) = f(\bot)$$

for all $x,y \in D$ and $f \in D$. They are obviously continuous. The map con matches every element of D with the corresponding *constant* function in $[D \to D]$. The map min associates to every function in $[D \to D]$ its *minimum* value in the partial ordering. The proof that this pair forms a projection is trivial. □

The pair con, min are not the only pair for making D a projection of $[D \to D]$. The following pair of maps were suggested by David Park:

$$\lambda x.\vec{e}[t,x] \quad \text{and} \quad \lambda f.f(t),$$

where x ranges over D, and f over $[D \to D]$ and where t is a fixed *isolated* element of D (that is, $t \prec t$ holds). The pair con and min will result if we set $t = \bot$. (Note that the expression $\vec{e}[t,x]$ though continuous in x is not continuous--or even monotonic--in the variable t.) A lattice may very well possess a large number of isolated elements, whence a large number of projections. And furthermore this is the only way the function $j = \lambda f.f(t)$ can be a projection. For assume the existence of an inverse $i : D \to [D \to D]$ satisfying the proper conditions. Then it would be the case that

$$i(x)(t) = x$$

and

$$i(f(t))(y) \sqsubseteq f(y)$$

for all $x,y \in D$ and all $f \in [D \to D]$. We can prove for all $U \in D$, if $t \not\sqsubseteq U$, then

$$i(x)(U) = \bot$$

by substituting $\vec{e}[v,x]$ for f in the second equation above, where v is chosen so that $v \prec t$ but not $v \prec u$. But then note that

$$i(x)(t) = \bigsqcup\{i(x)(u) : u \prec t\}.$$

If not $t \prec t$, then $u \prec t$ implies $t \not\sqsubseteq u$, which leads to absurdity. Hence t must be isolated, and, as we noted earlier, the function i is uniquely determined as being the one we already knew. Aside from these pairs of projections one could obtain others by combinations with automorphisms. I was unable to determine whether there are further pairs of an essentially different nature.

The topic of projections in these spaces is rather interesting since one has in a way more freedom in T_0-spaces (particularly in injective spaces) than in ordinary spaces for defining functions. As another example, consider the Cartesian square $D \times D$. Aside from the two obvious projections onto D, there is also the "diagonal" system given by the pair:

$$\lambda x.(x,x) \quad \text{and} \quad \lambda(x,y).x \sqcap y$$

We shall note in the next section how the choice of an initial projection effects the construction of an inverse limit.

The projections are not the only useful functions in $[D \to D] \to D$. As a final example of what can be done in function spaces we mention the fixed-point operator.

<u>3.14 Proposition</u>. *For a continuous lattice* D *there is a uniquely determined continuous mapping*

$$\mathrm{fix} : [D \to D] \to D$$

where for all $f \in [D \to D]$ *and* $x \in D$

$$f(\mathrm{fix}(f)) = \mathrm{fix}(f)$$

and whenever $f(x) = x$, *then*

$$\mathrm{fix}(f) \sqsubseteq x.$$

Proof: The proof of the existence of minimal fixed points in complete lattices is well known, as was mentioned in the proof of 2.10. To establish the continuity, it is sufficient to remark that since all functions $f \in [D \to D]$ are continuous, we have

$$\mathrm{fix}(f) = \bigsqcup_{n=0}^{\infty} f^n(\bot)$$

where $f^n(x) = f(f(\ldots f(x)\ldots))$ (n times). Thus fix is the point-wise lub of continuous functions on $[D \to D]$ and is thus itself continuous. □

4. INVERSE LIMITS. By an *inverse system* of spaces we understand as usual a sequence

$$\langle X_n, j_n \rangle_{n=0}^{\infty}$$

of T_0-spaces and continuous maps $j_n : X_{n+1} \to X_n$. The space X_∞, called the *inverse limit* of the sequence, is constructed in the familiar way as that subspace of the product space consisting of exactly those infinite sequences

$$x = \langle x_n \rangle_{n=0}^{\infty},$$

where for each n we have $x_n \in X_n$,

and

$$j_n(x_{n+1}) = x_n.$$

The space X_∞ is given the product topology, and the maps $j_{\infty n} : X_\infty \to X_n$ such that

$$j_{\infty n}(x) = x_n$$

are of course continous and satisfy the recursion equation:

$$j_{\infty n} = j_n \circ j_{\infty(n+1)}.$$

Besides this we have the expected extension property for any system of continuous maps

$$f_n : Y \to X_n$$

where for each n

$$f_n = j_n \circ f_{n+1}.$$

Because, we can define

$$f_\infty : Y \to X_\infty$$

by the equation

$$f_\infty(y) = \langle f_n(y)\rangle_{n=0}^{\infty}$$

for all $y \in Y$; whence

$$f_n = j_{\infty n} \circ f_\infty$$

holds. So much for a review of inverse limits. In this paper our interest will center on rather special inverse systems and their limits.

<u>4.1 Proposition</u>. *Let* $\langle D_n, j_n\rangle_{n=0}^{\infty}$ *be an inverse system of continuous lattices where each* $j_n : D_{n+1} \to D_n$ *is a projection. Then the inverse limit space* D_∞ *is also a continuous lattice.*

Proof: We need only show that D_∞ as a T_0-space is *injective*. So suppose $f_\infty : X \to D_\infty$ is given and $X \subseteq Y$. Define $f_n : X \to D_n$ by $f_n = j_{\infty n} \circ f_\infty$. Let $\bar{f}_n : Y \to D_n$ be the *maximal* extension of f_n according to 3.8. Now we can see the point of Lemma 3.9: by this construction we guarantee that $\bar{f}_n = j_n \circ \bar{f}_{n+1}$. Hence the required $\bar{f}_\infty : Y \to D_\infty$ exists. □

I do not know at the time of writing whether this theorem on inverse limits of continuous lattices extends to sequences where, say, the j_n are only retractions. Fortunately, sufficiently many projections exist to make this construction useful. Note that by reference to the product space construction of D_∞, its lattice ordering is given simply by the relation:

$$x \sqsubseteq y \quad \text{iff} \quad x_n \sqsubseteq y_n \text{ for all } n.$$

<u>4.2 Proposition</u>. *Let* $\langle D_n, j_n\rangle_{n=0}^{\infty}$ *and* D_∞ *be as in* 4.1. *Then the maps* $j_{\infty n} : D_\infty \to D_n$ *are projections.*

Proof: The projections $j_n : D_{n+1} \to D_n$, as we know, have their uniquely determined inverses $i_n : D_n \to D_{n+1}$. We can define $i_{n\infty} : D_n \to D_\infty$ by the equation:

$$i_{n\infty}(x) = \langle y_m\rangle_{m=0}^{\infty}$$

where

$$y_m = \begin{cases} j_m(y_{m+1}) & \text{if } m<n, \\ x & \text{if } m=n, \\ i_m(y_{m-1}) & \text{if } m>n. \end{cases}$$

The proof that $i_{n\infty}$ and $j_{\infty n}$ form a projection is now an easy computation. □

One should note also the recursion equation:

$$i_{n\infty} = i_{(n+1)\infty} \circ i_n .$$

These maps also make it possible to state this useful equation:

$$x = \bigsqcup_{n=0}^{\infty} i_{n\infty}(x_n),$$

where $x \in \mathrm{D}_\infty$. It is easy to check that this is a monotone lub, and so we can say each $x \in \mathrm{D}_\infty$ *is the limit* of its projections x_n. In fact, from what we know about projections, x_n is the best possible approximation to x in the space D_n.

<u>4.3 Corollary</u>. *Let the spaces be as in 4.1 and 4.2. Let* D' *be any complete lattice and suppose continuous functions* $f_n\colon \mathrm{D}_n \to \mathrm{D}'$ *are given so that* $f_n = f_{n+1} \circ i_n$. *Then we can define* $f_\infty\colon \mathrm{D}_\infty \to \mathrm{D}'$ *by the equation:*

$$f_\infty(x) = \bigsqcup_{n=0}^{\infty} f_n(x_n)$$

for $x \in \mathrm{D}_\infty$, *and we have* $f_n = f_\infty \circ i_{n\infty}$. □

The import of this last result is that within the category of complete lattices, the space D_∞ is not only the inverse limit of the D_n, but it is also the *direct limit*. (One system of spaces here uses the j_n as connecting maps, the other the i_n.) This is the algebraic result that lies at the heart of our main result about inverse limits of function spaces.

Turning to function spaces, let $\mathrm{D} = \mathrm{D}_0$ be a given continuous lattice. As we have seen in 3.13, there are many ways of making D_0 a projection of $\mathrm{D}_1 = [\mathrm{D}_0 \to \mathrm{D}_0]$. Choose one such given by a pair i_0, j_0. Define recursively:

$$\mathrm{D}_{n+1} = [\mathrm{D}_n \to \mathrm{D}_n]$$

and introduce the pairs i_{n+1}, j_{n+1} making D_{n+1} a projection of D_{n+2} by the method of 3.7. Specifically we shall have for $x \in \mathrm{D}_{n+1}$ and $x' \in \mathrm{D}_{n+2}$:

$$i_{n+1}(x) = i_n \circ x \circ j_n,$$

$$j_{n+1}(x') = j_n \circ x' \circ i_n.$$

Since these spaces are more than continuous lattices being function spaces, it is interesting to note that the maps i_n preserve function value as an algebraic operation as follows:

$$i_n(f(x)) = i_{n+1}(f)(i_n(x)),$$

where $x \in D_n$ and $f \in D_{n+1}$. Thus in passing to the limit space D_∞ something of functional application must also be preserved. The precise result shows that indeed D_∞ becomes its own function space.

4.4 Theorem. *The inverse limit D_∞ of the recursively defined sequence $\langle D_n, j_n \rangle_{n=0}^{\infty}$ of function spaces is not only a continuous lattice, but it is also homeomorphic to its own function space $[D_\infty \to D_\infty]$.*

Proof: We can write down directly the pair of maps i_∞, j_∞ that provide the homeomorphism:

$$i_\infty(x) = \bigsqcup_{n=0}^{\infty} (i_{n\infty} \circ x_{n+1} \circ j_{\infty n}),$$

$$j_\infty(f) = \bigsqcup_{n=0}^{\infty} i_{(n+1)\infty}(j_{\infty n} \circ f \circ i_{n\infty}).$$

Note that these formulae are simply generalizations at the limit for the formulae we used to define i_n, j_n in the first place. Thus it is not surprising that they would provide a projection of $[D_\infty \to D_\infty]$ upon D_∞. Indeed we can compute out $j_\infty(i_\infty(x))$, noting that all the lubs are monotone and that a double monotone limit can always be replaced by a single one in view of the continuity of the operations involved, obtaining

$$\begin{aligned} j_\infty(i_\infty(x)) &= \bigsqcup_{n=0}^{\infty} i_{(n+1)\infty}(j_{\infty n} \circ i_{n\infty} \circ x_{n+1} \circ j_{\infty n} \circ i_{n\infty}) \\ &= \bigsqcup_{n=0}^{\infty} i_{(n+1)\infty}(x_{n+1}) \\ &= x. \end{aligned}$$

In the converse order the calculation is only a bit more complicated. The idea is that since all the functions f are continuous and since the elements x are the limits of their approximations, then each f is actually completely determined by its sequence of *restrictions* $j_{\infty n} \circ f \circ i_{n\infty} \in D_{n+1}$. This simple idea can be made more precise with the aid of a lemma about D_∞, which allows us in certain cases to recognize projections from limits.

4.5 Lemma. *Suppose for each n we have* $u_{(n+1)} \in D_{n+1}$ *and we let:*

$$u = \bigsqcup_{n=0}^{\infty} i_{(n+1)\infty}(u_{(n+1)}).$$

Then if

$$j_{n+1}(u_{(n+2)}) = u_{(n+1)}$$

for each n, we can conclude that:

$$j_{\infty(n+1)}(u) = u_{(n+1)}.$$

Proof: If the sequence $u_{(n+1)}$ satisfies the recursion, then the limit defining u is monotonic. Therefore by continuity of projection it suffices to prove that

$$j_{\infty(n+1)}(i_{(m+1)\infty}(u_{(m+1)})) = u_{(n+1)}$$

for all $m \geq n$. This is obvious for $m = n$, and it can be readily proved by induction for larger m using the various recursion equations. (Properly speaking the induction is done on the quantity $(m - n)$ using both n and m as variables.) □

Proof of 4.4 concluded: The lemma applies at once to our calculation, for we find:

$$\begin{aligned} i_\infty(j_\infty(f)) &= \bigsqcup_{n=0}^{\infty} (i_{n\infty} \circ j_{\infty n} \circ f \circ i_{n\infty} \circ j_{\infty n}) \\ &= \bigsqcup_{n=0}^{\infty} (i_{n\infty} \circ j_{\infty n}) \circ f \circ \bigsqcup_{n=0}^{\infty} (i_{n\infty} \circ j_{\infty n}) \end{aligned}$$

Here we have just applied the continuity of f to be able to confine the lub on the right. But now by the remark following 4.2, we note the functional equation:

$$\mathrm{id} = \bigsqcup_{n=0}^{\infty} (i_{n\infty} \circ j_{\infty n}),$$

and the proof that i_∞ and j_∞ are inverse to one another is complete. complete. □

We can explain the idea of this proof in other terms using a suggestion made to me by F. W. Lawvere. Consider the category of continuous lattices and projections. In that category our D_∞ is, as we have remarked, both a *direct* and an *inverse* limit. Note too that with regard to projections $[D \to D']$ is a *functor*, for we can also define $[j \to j']$ when the maps are projections. In this language our particular inverse system is defined by the recursion:

$$D_{n+1} = [D_n \to D_n] \text{ and } j_{n+1} = [j_n \to j_n],$$

where D_0 and j_0 are given in advance. Now the function space construction is continuous in its two arguments turning an inverse limit on the right into an inverse limit and a *direct* limit on the *left* also into an *inverse* limit. A repeated limit can be made into a simple limit, so we can write:

$$\begin{aligned} D_\infty &= \underleftarrow{lim} \langle D_n, j_n \rangle_{n=0}^{\infty} \\ &= \underrightarrow{lim} \langle D_n, i_n \rangle_{n=0}^{\infty} \end{aligned}$$

and

$$\begin{aligned} [D_\infty \to D_\infty] &= \underleftarrow{lim} \langle [D_n \to D_n], [j_n \to j_n] \rangle_{n=0}^{\infty} \\ &= \underleftarrow{lim} \langle D_{n+1}, j_{n+1} \rangle_{n=0}^{\infty} \\ &= D_\infty \ (\textit{up to isomorphism}). \end{aligned}$$

A full checking of the details involved would not make the argument appreciably simpler over the more "element-by-element" argument I have presented. In fact, the proofs are actually the same. But thinking of the result in terms of properties of functors does seem to isolate very well the essential idea and to show how simple it is. One must only add here a note of caution: the proper choice of category must be done with care. Thus it seems to me that the use of projections rather than arbitrary continuous maps is necessary. Inasmuch as I have not checked all details in this form, I hope what I say is correct.

Since we have shown $[D_\infty \to D_\infty]$ to be homeomorphic to D_∞, we can begin to regard them as the same. In particular there ought to be some function space structure to transfer to D_∞. This can be done by defining functional application for any elements $x, y \in D_\infty$ by the equation:

$$x(y) = \bigsqcup_{n=0}^{\infty} i_{n\infty}(x_{n+1}(y_n)).$$

Similarly we can define λ-abstraction on continuous expressions:

$$\lambda x.[\ldots x\ldots] = j_\infty(\lambda x{:}D_\infty.[\ldots x\ldots]),$$

and in this way D_∞ becomes a model for the λ-calculus of Church and Curry. The model-theoretic and proof-theoretic aspects of this result will be explained in another paper (Scott (1972)).

Suppose we were to start with the least, non-trivial lattice $\mathbb{O} = \{\top,\bot\}$ for D_0. Now $D_1 = [\mathbb{O} \to \mathbb{O}]$ has exactly three elements and there are just two ways of defining a projection $j_0{:}D_1 \to D_0$. They are illustrated in the figure:

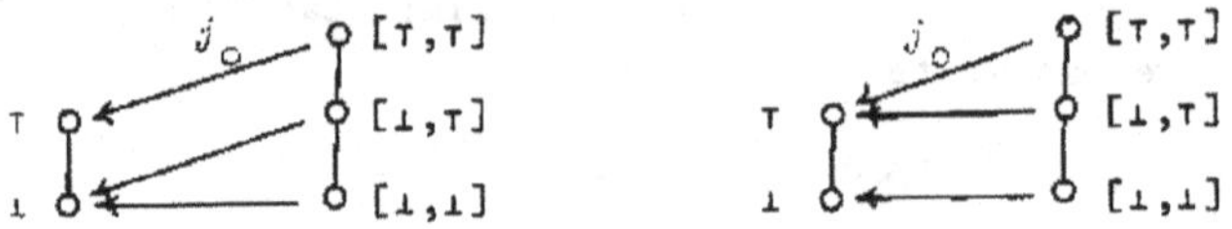

Hence our construction gives two limit spaces D_∞ and D'_∞. Are they the same? No, they are not. It can be shown, for example that the $\top$ of D_∞ is *isolated* (that is, $\top \prec \top$), while the same is not true of D'_∞. More interestingly, David Park has proved that the fixed-point operator fix mentioned in 3.14 has algebraic properties in D_∞ quite different from those in D'_∞. By *algebraic* here, we of course have reference to the functional algebra embodied in the application operation $x(y)$ defined on these limit spaces. Note, by the way, that in view of our isomorphism result we can regard fix (or any other similar continuous function for that matter) as an *element* of D_∞. This makes the "algebra" of D_∞ quite a rich field for study.

The reader will have surely remarked that by virtue of 1.5, every T_0-space X whatsoever can be embedded as a subspace in a D_∞. Besides this all the continuous functions on X (oh, into D_∞, say) can be extended to D_∞; whence they can be regarded as *elements* of D_∞. Thus we have been able to embed not only the topology of X into D_∞ but also all of the continuous function theory over X. So far this is only a "logical" construction. For more interesting "mathematical" results we shall have to investigate whether any useful theorems about the usual function spaces $[X \to X]$ can be obtained with the aid of D_∞. This method can easily be employed for real- or complex-valued continuous functions, though it seems more oriented toward pointwise convergence than anything else. Still, there seems to be a chance it might be useful--especially if one wished to consider continuous *operators* on function spaces.

The idea of forming the limit space can also be applied to *other functors* besides $[D \to D]$. Thus instead of solving the "equation"

$$D = [D \to D]$$

as we have done with the D_∞ construction, we could also solve:

$$V = T + [V \times V] + [V + V] + [V \to V]$$

for example. Here $T = \{\bot, 0, 1, \top\}$ is the four-element lattice with 0 and 1 as incomparable elements. By $[V \times V]$ and $[V \to V]$ we understand the usual Cartesian product and function space construction. The $+$ operator, on the other hand works only in the category of lattices with $\top$ as an isolated element. It is defined so as to make:

$$\begin{array}{ccc} D + D' & \longleftarrow & D' \\ \uparrow & & \uparrow \\ D & \longleftarrow & \mathbb{O} \end{array}$$

a push-out diagram, where the maps from $\mathbb{O}$ are meant to match $\bot$ with $\bot$ and $\top$ with $\top$. The point of requiring $\top$ to be isolated is that both D and D' become projections of $D+D'$. This construction, though not quite a disjoint union, has many properties in common with that operation on spaces. In particular, if we consider the category with projections as maps, the construction

$$\mathbb{F}(D) = T + [D \times D] + [D + D] + [D \to D]$$

is a *functor*. Furthermore, we can project $\mathbb{F}(T)$ onto T in an obvious way, thereby setting things up for an inverse limit construction:

$$V = \varprojlim \langle \mathbb{F}^n(T), j_n \rangle_{n=0}^{\infty}.$$

The resulting continuous lattice satisfies the desired equation up to isomorphism.

The space V constructed in the way just indicated is very rich in subspaces. To see this, consider the space J_V' of proper projections j where $j(\top) = \top$. As in 3.12 this is a complete lattice. Now that $[V \times V]$ and $[V + V]$ and $[V \to V]$ are regarded as *subspaces* of this "universe" V itself, we can easily define *continuous* operations

$$(j \times k)\ ,\ (j + k)\ ,\ \text{and}\ (j \to k)$$

on the projections obtaining again elements of J_V'. The projections so obtained correspond to the indicated constructions of subspaces, of course. (Indeed, if we had the time and space, we could show that J_V' becomes a very interesting category). There will be a particular projection t corresponding to T, and reason for doing all this is to show that the existence of subspaces of V can now be established by solving equations in $\mathcal{J}_V$. For example, by the fixed-point construction we could find a $j \in \mathcal{J}_V$ such that

$$j = t + (t\times j) + ((j\times j)\to j).$$

The range of j would then be a subspace $W \subseteq V$ such that W solves the equation:

$$W = T + [T\times W] + [[W\times W] \to W].$$

And these are only a few examples: simultaneous equations are possible, and many other operators are waiting for discovery and application.

REFERENCES. An announcement of this work and related investigations was first given in Scott (1970). Rather complete references and background material can be found in Scott (1971). A discussion of formal theories is to appear in Scott (1972).

The presentation of the material of the paper changed considerably after the January conference. In the first place remarks by several participants, Ernie Mannes in particular, caused me to rethink several points. Then the opportunity of lecturing at the Project MAC Seminar at MIT during the spring provided the opportunity of trying out some new ideas; these were then codified after lectures at the University of Southern California with the aid of several very helpful discussions on topology with James Dugundji.

The outcome of this development was that I found I could describe the work in purely topological terms in a simple and natural way leaving the lattices to be introduced as a special technique of analysis. This gives the presentation a much less *ad hoc* appearance, and relates the results to standard point-set topology in a much more understandable way. No doubt the whole idea of using completeness, inverse limits, and continuous functions could be put into a more general, more abstract categorical context, but I am not the man to do it. My interests at present lie in the direction of specific applications, though I can see that there might be some worthwhile directions to pursue.

For example, in understanding the connections of my kind of spaces with other topologies, one should consider the remarks on the topology of lattices in Birkhoff's paper in Abbott (1970). Some older papers such as Strother (1955) or Michael (1951) might also give some leads. It is curious how little there is of interest in the literature on T_0-spaces. Concerning function spaces there ought to be some connections with the *limit spaces* of Cook and Fischer (see especially Binz and Keller (1966)) and possibly with the notion of *quasi-topology* of Spanier (1963), but these are rather vague ideas.

In a different direction note that the *algebraic lattices* of Grätzer (1968) are in fact continuous lattices in which isolated points are dense. The continuous lattices may be "higher dimensional" while algebraic lattices are "zero dimensional" - in some suitable sense. Every continuous lattice is a retract of an algebraic lattice. But does this mean anything? Specific bibliographical references follow:

J. C. Abbott, ed., Trends in Lattice Theory, Van Nostrand Reinhold Mathematical Studies, vol.31 (1970).

P. Alexandroff and H. Hopf, Topologie I, Springer-Verlag, (1935).

E. Binz and H. H. Keller, *Funktionenräume in der Kategorie der Limesräume*, Annales Academiae Scientiarum Fennicae, Series A, I. Mathematica, no. 383 (1966), 21 pp.

G. Birkhoff, Lattice Theory, American Mathematical Society Colloquium Publications, vol. 25, Third (new) edition (1967).

E. Čech, Topological Spaces (revised by Z. Frolić and M. Katětov), Prague (1966).

G. Grätzer, Universal Algebra, Van Nostrand, (1968).

J. L. Kelley, General Topology, Van Nostrand, (1955).

E. Michael, *Topologies on Spaces of Subsets*, Transactions of the American Mathematical Society, vol. 77 (1951), pp. 152-182.

A. Nerode, *Some Stone Spaces and Recursion Theory*, Duke Mathematical Journal, vol. 26 (1959), pp. 397-406.

D. Scott, *Outline of a Mathematical Theory of Computation*, Proceedings of the Fourth Annual Princeton Conference on Information Sciences and Systems (1970), pp. 169-176.

________, *Lattice Theory, Data Types, and Semantices*, New York University Symposia in Areas of Current Interest in Computer Science (Randall Rustin ed.) (1971) to appear.

________, *Lattice-theoretic Models for Various Type-free Calculi*, Proceedings of the IVth International Congress for Logic, Methodology, and the Philosophy of Science, Bucharest (1972), to appear.

E. Spanier, *Quasi-topologies*, Duke Mathematical Journal, vol. 30 (1963) pp. 1-14.

W. Strother, *Fixed Points, Fixed Sets, and M-Retracts*, Duke Mathematical Journal, vol. 22 (1955), pp. 551-556.

Correction (Added March, 1972). Robin Milner has pointed out to me that there is an error in the remark in the paragraph immediately proceding Proposition 2.5. I was mistaken in saying that if D is a T_0-space which becomes a complete lattice under its induced partial ordering, then every set open in the given topology is also open in the induced topology. There are many counterexamples to this statement. Let D be any complete lattice. There are two extreme T_0-topologies which will induce the given partial ordering. The *smallest* such topology has as a sub-base for its open sets those sets of the form:

$$\{x \in D : x \not\sqsubseteq y\}.$$

These sets are easily proved to be open in any T_0-topology which induces the partial ordering. At the other extreme consider sets of the form:

$$\{x \in D : y \sqsubseteq x\}.$$

Such sets will give a base for a T_0-topology that is the *maximal* topology inducing the given partial ordering. Clearly they need not be open in the induced lattice topology; in particular, they may well fail to satisfy conditions (*ii*) on open sets. To make the remark in question correct, we must thus suppose that the given T_0-topology is *contained within* the induced lattice topology. The equation given in the paragraph indicated will then be a sufficient condition for the two topologies to be identical.

The remark was employed in the proof of three different propositions: 2.9, 2.10, and 3.3. In the case of 2.9 one must verify that the product topology is contained within the lattice topology. This need only be done for the basis for the product topology, and for such basic open sets the result needed is obvious. In the case of proposition 2.10 the question concerns a relationship between the topologies of a space and a subspace; the spaces in question are also lattices. Note in passing that a lub in the subspace is generally *larger* in the partial ordering than the corresponding lub relative to the whole space. This puts the inequalities in the wrong direction, and so it is not immediate that a relativized open set for the subspace is open in the lattice topology of the subspace. However, in this case we can appeal to the continuous retraction. Recall that the relativized open sets of the kind that we used in the proof of 2.10 are of the form:

$$U = \{z \in D : x < z\} .$$

Suppose then that S is a directed set, and that using the lub in the sense of D we have

$$\bigsqcup S \in U .$$

Referring back to the proof of 2.10 we know that

$$j(\bigsqcup{}' S) = \bigsqcup S ,$$

which means that

$$x < j(\bigsqcup{}'S) .$$

From this it follows that

$$x < j(z),\ \textit{some}\ z \in S\ .$$

Now $j(z) = z$, and we have what we need. This argument suffices only for a special type of open sets; but these open sets form a base for the topology, and so the argument is quite general.

Turning now to the proof of theorem 2.3 we note that the topology on the function space is simply the *relativized* product topology. There is no difficulty with lubs in this case, because, as we showed in the proof of that theorem, all lubs are calculated pointwise. Thus, it is easy to verify now that the sets open in the product topology are also open in the lattice topology.

SOME APPLICATIONS OF THE FORMALISM OF DUALITY IN ALGEBRAIC GEOMETRY

by

I. Bucur

§0. In the first part of my talk I'll recall some applications of the formalism of duality in the sense of GROTHENDIECK-VERDIER. The main results exposed here were obtained by VERDIER. Particularly we recall how it is possible to define in a very general context a LEFSCHETZ-Number $Lef(f,u)$, associated to the pair (f,u), where

$$f:X \longrightarrow X$$

is an endomorphism of the scheme X,

$$u:f^*(F) \longrightarrow F$$

is a "lifting" of f and F is a sheaf over X in a suitable topology subject to some finiteness conditions.

In fact VERDIER defines also a trace $Tr(f,u)$ associated to the pair (f,u) and proves a general LEFSCHETZ - fixed point formula in the case of étale topology. VERDIER shows also that $Lef(f,u)$ is a sum of local terms associated to the fixed points of f, but unfortunately it is not easy to find connections between these local terms introduced by VERDIER and other known invariants.

In the particular case of schemes of dimension one GROTHENDIECK introduced other local terms and conjectures some divisibility properties about them. We will prove in the second part of this talk some weaker divisibility properties of these local terms of GROTHENDIECK.

§1. Fibred categories. A fibred category F is given by

i) a category $\underset{\sim}{C}$, called the base of F

ii) a family $(F_X)_{X \in Ob(\underset{\sim}{C})}$ of categories indexed by the objects of the category $\underset{\sim}{C}$; the category F_X is called the fiber of F over X.

iii) a mapping α which assigns to every morphism $f:X \longrightarrow Y$ in C a functor $f^{\alpha}: F_Y \longrightarrow F_X$

iv) a mapping c which assigns to every pair (f,g) of composable morphisms of $\underset{\sim}{C}$,

$$X \xrightarrow{f} Y \xrightarrow{g} Z$$

an isomorphism $c_{f,g}$ of functors:

$$c_{f,g}:(gf)^{\alpha} \longrightarrow f^{\alpha}g^{\alpha} .$$

These data must satisfy the following conditions:

α) $(id)^{\alpha}$ = identity

β) $c_{f,id} = id_{f^{\alpha}}$, $c_{id,g} = id_{g^{\alpha}}$

for every three morphisms f,g,h in $\underset{\sim}{C}$:

$$X \xrightarrow{f} Y \xrightarrow{g} Z \xrightarrow{h} T$$

the following diagram is commutative:

$$\begin{array}{ccc} ((hg)f)^{\alpha} = (h(gf))^{\alpha} & \xrightarrow{c_{gf,h}} & (gf)^{\alpha}h^{\alpha} \\ {\scriptstyle c_{f,hg}} \downarrow & & \downarrow {\scriptstyle c_{f,g} \circ h^{\alpha}} \\ f^{\alpha}(hg)^{\alpha} & \xrightarrow{f^{\alpha} \circ c_{g,h}} & f^{\alpha}(g^{\alpha}h^{\alpha}) = (f^{\alpha}g^{\alpha})h^{\alpha} \end{array}$$

Dualising (taking the category $\underset{\sim}{C}^{\circ}$ instead of the category $\underset{\sim}{C}$) one obtains the notion of cofibred category.

1.1. Examples.1 Let $\underset{\sim}{C}$ be the category of ringed spaces and for every $(X,\mathcal{O}_X)\in Ob(\underset{\sim}{C})$ let be F_X the category $Mod(\mathcal{O}_X)$ of $\mathcal{O}_X$-modules. If $f:(X,\mathcal{O}_X)\longrightarrow(Y,\mathcal{O}_Y)$ is a morphism of ringed spaces we denote $f^*:F_Y\longrightarrow F_X$ the inverse image functor and if f,g are composable morphisms of ringed spaces, $c_{f,g}:(gf)^*\longrightarrow f^*g^*$ is the obvious isomorphism.

One obtains a fibred category having as base the category of ringed spaces.

2. Let $\underset{\sim}{C}$ be, as before, the category of ringed spaces and for every $(X,\mathcal{O}_X)\in Ob(\underset{\sim}{C})$ let be $F_X = D^+(Mod\ \mathcal{O}_X)$. (We denote by $D(A)$ the derived category of the abelian category A). If $f:(X,\mathcal{O}_X)\longrightarrow(Y,\mathcal{O}_Y)$ is a morphism of ringed spaces we denote by $\mathbb{R}f_*:D^+(Mod\ \mathcal{O}_X)\longrightarrow D^+(Mod\ \mathcal{O}_Y)$ the derived functor of the direct image functor associated with f. It is easy to see that for every pair of composable morphisms f,g of ringed spaces, there exists a canonical isomorphism:

$$c_{g,f}: \mathbb{R}(gf)_*\longrightarrow(\mathbb{R}g_*)(\mathbb{R}f_*)$$

In this way we obtain a cofibred category.

It is well to observe that in the first example, the fibers are abelian categories and the functors f^* are additive functors; in the second example the fibers are triangulated categories and the functors $\mathbb{R}f_*$ are exact functors.

§2. <u>Trace morphisms</u>. Let $F = (\underset{\sim}{C},(F_X)_{X\in Ob(\underset{\sim}{C})},(f\mapsto f^{\alpha}),((f,g)\to c_{f,g}))$ be a fibered category and $F' = (\underset{\sim}{C},(F'_X)_{X\in Ob(\underset{\sim}{C})},(f\longmapsto f_{\alpha}),((f,g)\longrightarrow c'_{f,g}))$ be a cofibred category with the same base as F and such that $F'_X \supseteq F_X$.

A <u>trace morphism</u> associated with the pair (F,F') is a mapping which assigns to every morphism

$$f:X\longrightarrow Y$$

of $\underset{\sim}{C}$, a natural morphism

$$Tr_f : f_\alpha f^\alpha \longrightarrow id_{F_Y}$$

such that the following conditions are satisfied:

(2.1) $\qquad Tr_{id} = \text{identity}$

(2.2) If f,g are composable morphisms in the category $\underset{\sim}{C}$,

$$X \xrightarrow{f} Y \xrightarrow{g} Z$$

then the following diagram

$$\begin{array}{ccc}
(gf)_\alpha (gf)^\alpha & \xrightarrow{Tr_{gf}} & id_{F_Z} \\
\downarrow {\scriptstyle (gf)_\alpha \circ c_{f,g}} & & \uparrow {\scriptstyle Tr_g} \\
(gf)_\alpha (f^\alpha g^\alpha) & & g_\alpha g^\alpha \\
\downarrow {\scriptstyle c'_{g,f} \circ f^\alpha g^\alpha} & & \uparrow {\scriptstyle g_\alpha \circ Tr_f \circ g^\alpha} \\
(g_\alpha f_\alpha)(f^\alpha g^\alpha) & = & g_\alpha (f_\alpha f^\alpha) g^\alpha
\end{array}$$

is commutative.

2.1. Example. Let k be an algebraicly closed field, X a connected normal curve, proper over k and K its function field. We denote by $\underset{\sim}{C}$ the sub-category of the category of algebraic varieties defined over k , generated by the curve X and the final element $*$. $*$ is the ringed space reduced to a single point with structural sheaf k . The single non-identity morphism in C is the canonical projection

$$p : X \longrightarrow *$$

It is easy to define a cofibred category F having as base the category $\underset{\sim}{C}$.

We will put:

— $F_X = \mathcal{P}ar\mathcal{f}(X)$ = the sub category of $D(\text{Mod } \mathcal{O}_X)$ generated by complexes of $\mathcal{O}_X$-modules with coherent cohomology and of finite global tor-dimension.

— $F_* = \mathcal{P}ar\mathcal{f}(k)$.

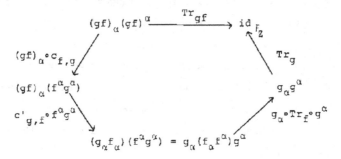

The functor associated with p will be $\mathbb{R}\, p_*: \mathcal{Parf}(X) \longrightarrow \mathcal{Parf}(k)$

It is possible to define also a fibered category $\mathcal{G}$ with the same base and the same fibres as $\mathcal{F}$, putting:

(2.1.1) $\quad p^! \; : \; \mathcal{Parf}(k) \longrightarrow \mathcal{Parf}(X)$,

$$p^!(k) = \Omega_{X/k}[1]$$

where, $\Omega_{X/k}$ is the sheaf of 1-differential forms on X defined over k .

We want to define the natural morphism

(2.1.2) $\quad Tr_p : \mathbb{R}p_* \circ p^! \longrightarrow id_{\mathcal{F}_*}$

It is enough to define a morphism

$$Tr_p(k) : (\mathbb{R}\, p_* \circ p^!)(k) \longrightarrow k$$

in the category D(k) . But every complex of vectorial spaces over k

$$L = (L_i, d_i^L)_{i \in \mathbf{Z}} \quad ,$$

bounded at the left or at the right, is isomorphis with the complex:

$$(\ldots \longrightarrow H^i(L) \xrightarrow{\;0\;} H^{i+1}(L) \longrightarrow \ldots)$$

whose differentials are all zero. Consequently the complex $(\mathbb{R}\, p_* \circ p^!)(k)$ is isomorphic (in D(k)) with the complex:

$$(\ldots \longrightarrow \circ \longrightarrow \overset{-1}{H^0(X,\Omega_{X/k})} \longrightarrow \overset{0}{H^1(X,\Omega_{X/k})} \longrightarrow \circ \longrightarrow \ldots)$$

Therefore it must define a morphism of vertor spaces:

(2.1.3) $\quad Tr_p(k) : H^1(X,\Omega_{X/k}) \longrightarrow k$

The vector space $H^1(X,\Omega_{X/k})$ is canonically isomorphic (cf.J.P. Serre, Groupes algébriques et corps de classes,...) with

$$R/_{(R(\Omega_{X/k})+K)}$$

where:

R is the algebra of distributions over X , i.e. the algebra of systems $(f_x)_{x \in X}$, $f_x \in K$ and $f_x \in \mathcal{O}_x$ for almost all $x \in X$.

Let ω_0 be a meromorphic differential form over X . $R(\Omega_{X/k})$ will be by definition the sub-space of R formed by the distributions $(f_x)_{x\in X}$ such that

(2.1.4) $\quad v_x(f_x\omega_0) = v_x(f_x) + v_x(\omega_0) \geq 0$

(v_x is the valuation of K defined by the point x of X)
Of course K can be immersed in R by

$$f \longrightarrow (f_x)_{x\in X} \quad , \quad f_x = f \quad \forall x\in X$$

Using the notion of residue of a meromorphic differential form, one defines Tr_p in the following manner:

(2.1.5) $\quad Tr_p(\widehat{(f_x)}) = \sum_{x\in X} res(f_x\omega_0) \quad , \quad \forall (f_x)_{x\in X}\in R$.

The condition (2.1.4) says that Tr_p vanishes on $R(\Omega_{X/k})$ and "the theorem of residues" says that it vanishes on K , therefore the morphism (2.1.3) is well defined by (2.1.5)

§3. <u>Duality morphisms</u>. Let $F = (\underset{\sim}{C}, (F_X)_{X\in Ob(\underset{\sim}{C})}, (f\longrightarrow f^{\alpha}), c)$ be a fibered category whose fibers are sub-categories of derived categories i.e. $F_X \subseteq D(A_X)$, where A_X is an abelian category and such that for every morphism

(3.1) $$f: X \longrightarrow Y$$

in $\underset{\sim}{C}$, the functor

$$f^{\alpha}: F_Y \longrightarrow F_X$$

is exact.

If $F' = (\underset{\sim}{C}, (F_X)_{X\in Ob(\underset{\sim}{C})}, (f \longmapsto f_{\alpha}), c')$ is a cofibered category with the same base as F , $F'_X \supseteq F_X$, $(f \longmapsto Tr_f)$ a trace morphism associated with the pair (F, F') and if for every morphism (3.1) in $\underset{\sim}{C}$, the functor

$$f_{\alpha}: F_X \longrightarrow F_Y$$

is exact, then it is possible to define for every morphism (3.1), every $i \in \mathbb{Z}$ and every pair (A,B), $A \in Ob(F_X)$, $B \in Ob(F_Y)$, a natural morphism

$$(3.2) \qquad \theta_f^i : \mathrm{Ext}^i_{A_X}(A, f^\alpha(B)) \longrightarrow \mathrm{Ext}^i_{A_Y}(f_\alpha(A), B) \quad ,$$

called duality morphisms, composing the two, following morphisms:

$$\mathrm{Ext}^i_{A_X}(A, f^\alpha(B)) = \mathrm{Hom}_{D(A_X)}(A, T^i(f^\alpha(B))) \xrightarrow{f_\alpha(A, T^i(f^\alpha(B)))}$$

$$\mathrm{Hom}_{D(A_Y)}(f_\alpha(A), T^i(f_\alpha f^\alpha(B)))$$

$$\mathrm{Hom}_{D(A_Y)}(f_\alpha(A), T^i(f_\alpha f^\alpha(B))) \xrightarrow{\mathrm{Hom}_{D(A_Y)}(f_\alpha(A), Tr_f(B))}$$

$$\mathrm{Hom}_{D(A_Y)}(f_\alpha(A), T^i(B)) = \mathrm{Ext}^i_{A_Y}(f_\alpha(A), B)$$

§4. A particular case.

The morphisms (3.2) can be specified in the particular case when the category $\underset{\sim}{C}$ is a sub-category S of the category of ringed sites and when for every $X = (S, A) \in Ob(\underset{\sim}{C})$, F_X is a subcategory of $D(\mathrm{Mod}\ A)$.

Let $f : X = (S, A) \longrightarrow (S', A") = Y$ be a morphism of S. Every injective A-module is acyclic for the functor f_* and every flat A'-,module is acyclic for the functor f^*.

Consequently there exist the derived functors:

$$(4.1) \qquad \mathbb{R} f_* : D^+(\mathrm{Mod}\ A) \longrightarrow D^+(\mathrm{Mod}\ A')$$

$$(4.2) \qquad \mathbb{L} f^* : D^-(\mathrm{Mod}\ A') \longrightarrow D^-(\mathrm{Mod}\ A) \ .$$

In the case when $\mathbb{R} f_*$ (resp. $\mathbb{L} f^*$) is of finite cohomological dimension it is possible to omit the exponent $+$(resp.$-$).

With the same notations as above we assume that $\underset{\sim}{C}$ is a subcategory S of the category of ringed sites, for every $X = (S, A) \in Ob(\underset{\sim}{C},)$, F_X(resp F_Y) is a subcategory of $D(\mathrm{Mod}\ A)$, and $F'_X \supseteq F_X$. Then for every $f:X \longrightarrow Y$ and every pair (A,B), $A \in \mathcal{O}b(F_X) \cap \mathcal{O}b(D^-(\text{Mod } A))$, $B \in \mathcal{O}b(F_Y) \cap \mathcal{O}b(D^+(\text{Mod } A'))$ it is possible to define a natural morphism

$$(4.3) \qquad \theta_f : \mathbb{R} f_* \mathbb{R}\mathcal{H}om_X(A, f^\alpha(B)) \longrightarrow \mathbb{R}\mathcal{H}om_Y(f_\alpha(A), B)$$

called <u>morphism of duality</u> by the composition of the two following morphisms:

$$(4.4) \qquad \mathbb{R} f_* \mathbb{R}\mathcal{H}om_X(A, f^\alpha(B)) \longrightarrow \mathbb{R}\mathcal{H}om_Y(f_\alpha(A), f_\alpha(f^\alpha(B)))$$

$$(4.5) \qquad \mathbb{R}\mathcal{H}om_Y(f_\alpha(A), (f_\alpha f^\alpha)(B)) \xrightarrow{Tr_f} \mathbb{R}\mathcal{H}om_Y(f_\alpha(A), B)$$

When we apply the functor global sections on Y we obtain the functorial morphism

$$(4.6) \qquad \mathbb{R}\,\mathrm{Hom}_X(A, f^\alpha(B)) \longrightarrow \mathbb{R}\,\mathrm{Hom}_Y(f_\alpha(A), B)$$

which yields, taking the cohomology , the functorial morphisms of groups (3.2).

We assume now that the category $\underset{\sim}{C}$ has a final element $e = (*, \mathcal{O}_*)$. If π_X is the canonical morphism of X on the final element and $\mathcal{O}_* \in \mathcal{O}b(F_e)$, we denote

$$K_X = \pi_X^\alpha(\mathcal{O}_*)$$

$$\underline{D}_X = \mathbb{R}\mathcal{H}om(.,K_X) \quad .$$

With these notations, the duality morphism θ_f gives the morphism

$$\mathbb{R} f_*(\underline{D}_X(A)) \longrightarrow \underline{D}_Y(f_\alpha(A))$$

<u>A theory of duality in the sense of Grothendieck is a pair of categories (F,F') as above and a trace morphism such that the associated duality morphisms are isomorphisms.</u>
<u>To state a theorem of duality in the sense of Grothendieck, means to find conditions which imply the fact that the duality morphism θ_f is an isomorphism.</u>

(4.7) Example. a) (Grothendieck, Hartshorne,....) Let $F = (\underset{\sim}{C}, (F_X)_{X\varepsilon Ob(\underset{\sim}{C})}, (f \longrightarrow \mathbb{R}f_*), c_{f,g})$ be the cofibered category defined as follows:

— $\underset{\sim}{C}$ is the category of noetherian schemes of finite Krull dimensions and whose morphisms are proper morphisms of schemes.

— $F_X = D^+_{qc}(X)$ = the subcategory of $D(\text{Mod } \mathcal{O}_X)$, generated by complexes of arbitrary $\mathcal{O}_X$-modules, bounded at the left and whose cohomology sheaves are quasi-coherent

— $\mathbb{R}f_*$ is, of course, the direct image functor.

There exists a fibered category $F^! = (C, (F_{X\varepsilon Ob(\underset{\sim}{C})}, (f \longrightarrow f^!), c^!_{f,g})$ with the same base and fibers as $F^!$ and a trace morphism associated to the pair $(F, F^!)$, such that for any $F \varepsilon D_{qc}(X)$, $G\varepsilon D^+_{qc}(Y)$ and any morphism

$$f : X \longrightarrow Y$$

in $\underset{\sim}{C}$, the duality morphism:

$$\theta_f : \mathbb{R}f_* \mathbb{R}Hom_X(F, f^!(G)) \longrightarrow \mathbb{R}Hom_Y(\mathbb{R}f_*(F), G)$$

is an isomorphism. Moreover:

i) if f is a smooth morphism of relative dimension d, then

(4.7.1) $$f^!(G) = \mathbb{L}f^*(G) \otimes \omega_{X/Y}[d]$$

where $\omega_{X/Y}$ is the sheaf of highest differential.

ii) if f is a finite morphism then

(4.7.2) $$f^!(G) = \bar{f}^* \mathbb{R}Hom_Y(f_* \mathcal{O}_X, G)$$

where $\bar{f}$ is the morphism of ringed spaces $(X, \mathcal{O}_X) \longrightarrow (Y, f_*(\mathcal{O}_X))$.

b) (Grothendieck, Verdier, Deligne,...). Let $F = (\underset{\sim}{C}, (F_X)_{X\varepsilon Ob(\underset{\sim}{C})}, (f \longmapsto \mathbb{R}f_*), c_{f,g})$ be the cofibred category defined as follows:

$\underset{\sim}{C}$ is the category of locally noetherian schemes defined over an algebraically closed field k of characteristic p, and whose morphisms are proper morphisms of schemes defined over k.

— $F_X = D^+_n(X)$, where $D_n(X)$ is defined in the following manner: Let n be prime with p and $(\mathbb{Z}/_{n\mathbb{Z}})_X$ the sheaf over X in the étale topology, associated to the ring $\mathbb{Z}/n\mathbb{Z}$. $D_n(X)$ will be $D(\mathrm{Mod}(\mathbb{Z}/n\mathbb{Z})_X)$. $\mathbb{R}f_*$ is as before, the direct image functor. There exist a fibered category $F^! = (\underset{\sim}{C}, (F_X)_{X\in Ob(\underset{\sim}{C})}, (f \longmapsto f^!), c^!_{f,g})$, with the same base and fibers as F and a trace morphism associated to the pair $(F,F^!)$ such that for any $F\in D_n^-(X), G\in D_n^+(Y)$ and any morphism

$$f:X \longrightarrow Y$$

in $\underset{\sim}{C}$, the duality morphism θ_f is an isomorphism. Moreover if f is a smooth morphism of relative dimension d , then:

$$(4.7.3) \qquad f^!(G) = \mathbb{L}f^*(G)\otimes T_{X/Y}$$

where $T_{X/Y}$, called the <u>orientation complex</u>, is defined as follows.

Let μ_n be the sheaf of nth roots of unity over X . Then

$$(4.7.4) \qquad (T_{X/Y})_i = \begin{cases} 0 & \text{if } i \neq -2d \\ \mu_n\otimes\mu_n\otimes\cdots\otimes\mu_n = \mu_n^{\otimes d} & \text{if } i = 2d \end{cases} .$$

§5. <u>Applications. Iterate residues</u> . Let $f:X \longrightarrow Y$ be a proper and smooth morphism of noetherian schemes of relative dimension n , and $i:Z \hookrightarrow X$ a closed subscheme of X , finite over Y and defined locally by an $\mathcal{O}_X$-sequence. Choosing $t_1,t_2,\ldots,t_n$ an $\mathcal{O}_X$-sequence of parameters which generate the ideal of Z locally around a closed point $z_0\in Z$ and ω a differential form of degree n on X defined in a neighbourhood of z_0 we can define a <u>residue symbol</u>

$$\mathrm{Res}_{z_0}\left[{\omega \atop t_1,t_2,\ldots,t_n}\right]\in\Gamma(Y,\mathcal{O}_Y)$$

as follows. Let $g:Z \longrightarrow Y$ be the composed morphism fi . Using the notations of 4.7, (4.7.1) and (4.7.2) and the trace morphism $\mathrm{Tr}_g(\mathcal{O}_Y):\mathbb{R}g_*g^!(\mathcal{O}_Y) \longrightarrow \mathcal{O}_Y$ we obtain the morphisms

$$(5.1) \qquad g^!(\mathcal{O}_Y) \simeq i^!(f^!(\mathcal{O}_Y)) = i^!(\omega_{X/Y}[n])$$

(5.2) $\mathbb{R}\, i_* i^! (\omega_{X/Y}[n]) \simeq \mathbb{R}\,\mathcal{H}om_X(\mathcal{O}_Z, \omega_{X/Y}[n])$

(5.3) $\mathrm{Res}_Z : \mathrm{Ext}^n_X(\mathcal{O}_Z, \omega_{X/Y}) \longrightarrow \Gamma(Y, \mathcal{O}_Y)$.

The Koszul complex $K_*(t_1, t_2, \ldots, t_n); \mathcal{O}_{z_0})$ is a resolution of $\mathcal{O}_{Z,z_0}$:

(5.4) $\ldots \longrightarrow 0 \longrightarrow K_n((t_1, t_2, \ldots, t_n); \mathcal{O}_{z_0}) \longrightarrow \ldots \longrightarrow K_0((t_1, \ldots, t_n); \mathcal{O}_{z_0})$

$\longrightarrow \mathcal{O}_{Z,z_0} \longrightarrow 0$,

and we have:

(5.5) $K_n((t_1, t_2, \ldots, t_n); \mathcal{O}_{z_0}) = \Lambda^n(\mathcal{O}^n_{z_0})$.

Since $\Lambda^n(\mathcal{O}^n_{z_0})$ is free of rank one and has a canonical base, the differential form ω defines an element of $\mathrm{Hom}_{\mathcal{O}_{z_0}}(K_n((t_1, t_2, \ldots$ $\ldots, t_n); \mathcal{O}_{z_0}), \omega_{X/Y})$. One obtains in this way an element of $\mathrm{Ext}^n_{z_0}(\mathcal{O}_{z_0}; \mathcal{O}_{Z,z_0}, \omega_{X/Y,z_0})$ and finally an element

(5,6) $\left[\begin{matrix} \omega \\ t_1, t_2, \ldots, t_n \end{matrix}\right] \in \mathrm{Ext}^n_X(\mathcal{O}_Z, \omega_{X,Y})$.

Applying the residue map (4.3) we get the element $\mathrm{Res}_Z \left[\begin{matrix} \omega \\ t_1, t_2, \ldots, t_n \end{matrix}\right]$ denoted usually simply by $\mathrm{Res}_{z_0}\left[\begin{matrix} \omega \\ t_1, t_2, \ldots, t_n \end{matrix}\right]$.

§6. Dualizing complexes. Let $X = (S, A)$ be a ringed topos and $K \in$, $D^+(X) = D^+(\mathrm{Mod}\ A)$

For every $F \in (D(X))$ there is a natural functorial homomorphism:

(6.1) $\eta : F \longrightarrow \mathbb{R}\,\mathcal{H}om(\mathbb{R}\,\mathcal{H}om(F, K), K)$.

Indeed, we can assume first that the components of K are injectives and consequently we can erase the $\mathbb{R}$. Using now the isomorphism:

(6.2) $\mathrm{Hom}(\mathcal{H}om(F,K), \mathcal{H}om(F,K)) \xrightarrow{\sim} \mathrm{Hom}(F \otimes \mathcal{H}om(F,K), K) \xrightarrow{\sim} \mathrm{Hom}(F, \mathcal{H}om(\mathcal{H}om(F,K), K))$,

we obtain the homomorphism η as the image of the identity homomorphism of $\mathcal{H}om(F,K)$ by the homomorphism obtained by composition of the two homomorphisms (6.2).

We say that the complex $F \in D(X)$ is <u>reflexive</u> with respect to K if the natural homomorphism η is an isomorphism.

<u>Definition (6.5).</u> Let G_X be a subcategory of $D(X)$. A complex $K \in \mathcal{O}b(D^+(X))$ is called a <u>dualizing complex</u> for X with respect to G_X if every $F \in \mathcal{O}b(G_X)$ is reflexive with respect to K .

§7. <u>Lefschetz Number</u>. We assume the hypotheses and notations of §4

Let $X = (S,A)$ be a ringed site which belongs to $\mathcal{O}b(\underset{\sim}{C})$,

$F \in \mathcal{O}b(F_X) \cap \mathcal{O}b(F'_X)$, $f: X \longrightarrow X$ and $u: f^*(F) \longrightarrow F$ a "lifting" of f:

We will show that the formalism of duality in the sense of Grothendieck, i.e. the duality morphisms are isomorphisms, allows to define under suitable conditions, explained latter, the "<u>Lefschetz</u> <u>Number</u>" $Lef(f,u)$ of the pair (f,u) , as an element of $A_* = \Gamma(*,\mathcal{O}_*)$, $(*,\mathcal{O}_*)$ being the final element of the category $\underset{\sim}{C}$:

$$\text{(7.1)} \qquad Lef(f,u) \in \Gamma(*,\mathcal{O}_*) = H^\circ(e,\mathcal{O}_*) \quad ,$$

Under some finiteness assumptions on F is also possible to define the trace $Tr_{A_*}(f,u)$ as an element of A_*. Consequently one obtains a conjecturally <u>Lefschetz fixed point formula</u>:

$$\text{(7.2)} \qquad Tr_{A_*}(f,u) = Lef(f,u)$$

This formula is already proved in some particular (but very important) cases (Grothendieck, Verdier, Artin,...)

To define the "Lefschetz Number" $Lef(f,u)$ we need the following lemmas:

<u>Lemma 7.3</u>. We assume the category $\underset{\sim}{C}$ has finite products and for every morphism

$$f: U \longrightarrow V \quad ,$$

in $\underset{\sim}{C}$, $f_\alpha = \mathbb{R}f_*$. Then it is possible to associate canonically to the pair (f,u) two elements:

(7.3.1) $$c(\Delta)\in H^{\circ}(X\times X, \mathbb{R}\mathcal{H}om(pr_1^*(F), pr_2^{\alpha}(F))$$ (Δ denotes the diagonal of X×X)

$$c(f,u)\in H^{\circ}(X\times X, \mathbb{R}\mathcal{H}om(pr_2^*(F), pr_1^{\alpha}(F))$$

where $pr_1, pr_2 : X\times X \longrightarrow X$ are the canonical projections.

<u>Proof</u>. Since the definition of $c(\Delta)$ is similar and even easier than that of $c(f,u)$, we will give only the definition of $c(f,u)$

We consider the commutative diagram:

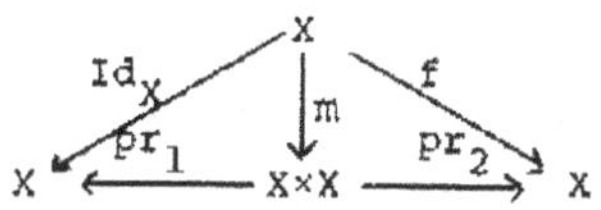

where m is the graph of f .

We have the canonical isomorphism:

(7.3.3) $$\mathbb{R}\, m_*\mathbb{R}\mathcal{H}om(f^*(F),F) \simeq \mathbb{R}\, m_*(m^*.pr_2^*(F)\ ,\ m^{\alpha}pr_1^{\alpha}(F))$$

(7.3.4) $$\mathbb{R}\, m_*\mathbb{R}\,\mathcal{H}om(m^*pr_2^*(F), m^{\alpha}pr_1^{\alpha}(F) \simeq \mathbb{R}\,\mathcal{H}om(\mathbb{R}\, m_*m^*pr_2^*(F), pr_1^{\alpha}(F)).$$

The morphism of adjointness $Id \longrightarrow \mathbb{R}\, m_*m^*$, gives us the , morphism

(7.3.5) $$\mathbb{R}\,\mathcal{H}om(\mathbb{R}\, m_*m^*pr_2^*(F), pr_1^{\alpha}(F)) \longrightarrow \mathbb{R}\mathcal{H}om(pr_2^*(F), pr_1^{\alpha}(F))$$

Finally one obtains the morphism:

(7.3.6) $$\mathcal{H}om(f^*(F),F) \longrightarrow H^{\circ}(X\times X, \mathbb{R}\,\mathcal{H}om(pr_2^*(F), pr_1^{\alpha}(F))\ .$$

The element $c(f,u)$ is by definition the image of u by the morphism (7.3.6) . We assume in the sequel the following supplementary hypotheses on the duality theory. This new hypotheses will be called the "<u>Künneth conditions</u>". Namely: With the notations of the following diagram

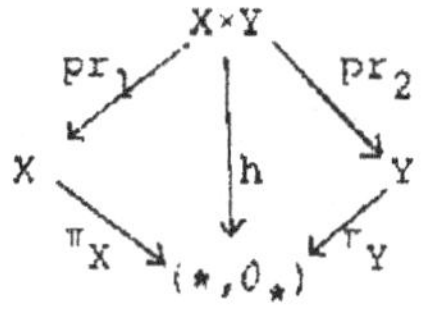

$h = \pi_{X\times Y}$

i) the "Künneth morphism"

(7.3.7) $$\mathbb{R}\,\pi_{X*}(K_X)\overset{\mathbb{L}}{\otimes}_{\mathcal{O}_*}\mathbb{R}\,\pi_{Y*}(K_Y) \xrightarrow{\;K\;} \mathbb{R}\,h_*(pr_1^*(K_X)\overset{\mathbb{L}}{\otimes} pr_2^*(K_Y))$$

is an isomorphism

ii) The morphism

(7.3.8) $$pr_1^*(K_X)\overset{\mathbb{L}}{\otimes} pr_2^*(K_Y) \longrightarrow K_{X\times Y}$$

canonically associated to K , is also an isomorphism. For defining (8.3.8) it is enough to define a morphism

(7.3.9) $$\mathbb{R}\,h_*(pr_1^*(K_X)\overset{\mathbb{L}}{\otimes} pr_2^*(K_Y)) \longrightarrow \mathcal{O}_*$$

Such a morphism may be defined using the isomorphism K given by the tensor product of the following two morphisms:

$$Tr_{\pi_X}(\mathcal{O}_*):\mathbb{R}\,\pi_{X*}(K_X)\longrightarrow \mathcal{O}_* \quad , \quad Tr_{\pi_Y}(\mathcal{O}_*):\mathbb{R}\,\pi_{Y*}(K_Y)\longrightarrow \mathcal{O}_* \quad .$$

Let be $F\in Ob(D(X))$, $G\in Ob(D(Y))$. It is also possible to associate canonically to the isomorphism (7.3.8) a morphism:

(7.3.10) $$pr_1^*\underline{D}_X(F)\overset{\mathbb{L}}{\otimes} pr_2^*\underline{D}_Y(G) \longrightarrow \underline{D}_{X\times Y}(pr_1^*(F)\overset{\mathbb{L}}{\otimes} pr_2^*(G)) \quad .$$

in the following manner:

We have obviously the sequence of morphisms:

$$pr_1^*\underline{D}_X(F)\otimes pr_2^*\underline{D}_Y(G) \longrightarrow \mathbb{R}\mathcal{H}om(pr_1^*(F),pr_1^*(K_X))\overset{\mathbb{L}}{\otimes}\mathbb{R}\mathcal{H}om(pr_2^*(G),pr_2^*(K_Y)) \quad .$$

$$\mathbb{R}\mathcal{H}om(pr_1^*(F),pr_1^*(K_X))\overset{\mathbb{L}}{\otimes}\mathbb{R}\mathcal{H}om(pr_2^*(G),pr_2^*(K_Y)) \longrightarrow$$

$$\mathbb{R}\mathcal{H}om(pr_1^*(F)\overset{\mathbb{L}}{\otimes} pr_2^*(G),pr_1^*(K_X)\overset{\mathbb{L}}{\otimes} pr_2^*(K_Y))$$

$$\mathbb{R}\mathcal{H}om(pr_1^*(F)\overset{\mathbb{L}}{\otimes} pr_2^*(G),pr_1^*(K_X)\overset{\mathbb{L}}{\otimes} pr_2^*(K_Y)) \longrightarrow \underline{D}_{X\times Y}(pr_1^*(F)\overset{\mathbb{L}}{\otimes} pr_2^*(G))$$

The last morphism is induced by the morphism 7.3.8. One obtains the morphism (7.3.10) by the composition of these morphisms.

<u>Lemma 7.4</u>. (Cf S.G.A.4, Exp.2 Corollary 3.2) Let be $F\in Ob(D(X))$, $G\in Ob(D(Y))$ such that the following conditions are satisfied:

a) The morphism (7.3.10) is an isomorphism.

b) G (resp. $pr_2^{\alpha}(G)$) is reflexive with respect to K_Y (resp. $K_{X\times Y}$)

c) F (resp. $pr_1^{\alpha}(F)$) is reflexive with respect to K_X (resp. $K_{X\times Y}$)

Under these conditions,

i) There exists a canonical isomorphism:

(7.4.1) $$\mathbb{R}\,\mathcal{H}om(pr_1^*(F), pr_2^{\alpha}(G)) \xrightarrow{\sim} pr_1^*(\underline{D}_X(F)) \overset{\mathbb{L}}{\otimes} pr_2^*(G) \quad .$$

ii) There exists a canonical pairing:

(7.4.2) $$\mathbb{R}\,\mathcal{H}om(pr_1^*(F), pr_2^{\alpha}(G)) \overset{\mathbb{L}}{\otimes} \mathbb{R}\,\mathcal{H}om(pr_2^*(G), pr_1^{\alpha}(F)) \longrightarrow K_{X\times Y} \quad .$$

<u>Proof</u>. Using the reflexivity of $pr_2^{\alpha}(G)$ with respect to $K_{X\times Y}$ the "Kunneth conditions" and the hypotheses a), the reflexivity of G with respect to K_Y . we obtain the sequence of isomorphisms:

$$\mathbb{R}\,\mathcal{H}om(pr_1^*(F), pr_2^{\alpha}(G)) \simeq \underline{D}_{X\times Y}(pr_1^*(F) \overset{\mathbb{L}}{\otimes} \underline{D}_{X\times Y}(pr_2^{\alpha}(G))$$

$$\underline{D}_{X\times Y}(pr_1^*(F) \overset{\mathbb{L}}{\otimes} \underline{D}_{X\times Y}(pr_2^{\alpha}(G))) \simeq pr_1^*(\underline{D}_X(F)) \overset{\mathbb{L}}{\otimes} pr_2^*(\underline{D}_Y(\underline{D}_Y(G)))$$

$$pr_1^*(\underline{D}_X(F)) \overset{\mathbb{L}}{\otimes} pr_2^*(\underline{D}_Y(\underline{D}_Y(G))) \simeq pr_1^*(\underline{D}_X(F)) \overset{\mathbb{L}}{\otimes} pr_2^*(G) \quad ,$$

which proves i) .

Using i) it is enough to define a morphism

(7.4.3) $$(pr_1^*(\underline{D}_X(F)) \overset{\mathbb{L}}{\otimes} pr_2^*(G)) \overset{\mathbb{L}}{\otimes} (pr_2^*(\underline{D}_Y(G)) \otimes pr_1^*(G)) \longrightarrow K_{X\times Y} \quad .$$

But we have obviously the canonical isomorphism:

$$pr_1^*(\underline{D}_X(F)) \overset{\mathbb{L}}{\otimes} pr_2^*(G) \overset{\mathbb{L}}{\otimes} (pr_2^*(\underline{D}_Y(G)) \overset{\mathbb{L}}{\otimes} pr_1^*(F)) \simeq pr_1^*(\underline{D}_X(F) \overset{\mathbb{L}}{\otimes} F) \overset{\mathbb{L}}{\otimes} pr_2^*(\underline{D}_Y(G) \overset{\mathbb{L}}{\otimes} G)$$

Using the biduality isomorphisms, one obtains the morphisms

$$pr_1^*(\underline{D}_X(F) \overset{\mathbb{L}}{\otimes} F) \longrightarrow pr_1^*(K_X) \quad ,$$

$$\mathrm{pr}_2^*(\underline{D}_Y(G)\overset{\mathbb{L}}{\otimes}G) \longrightarrow \mathrm{pr}_2^*(K_Y)$$

and consequently, using the isomorphism 7.3.8 the morphism (7.4.3)

Corollary 7.5. We assume the hypotheses of the lemmas 7.3,7.4 There exist a canonical pairing:

$$(7.5.1)\quad \mathrm{Lef}(F,G):H^\circ(X\times Y,\mathbb{R}\,\mathcal{H}om(\mathrm{pr}_1^*(F),\mathrm{pr}_2^\alpha(G)))\times H^\circ(X\times Y,\mathbb{R}\,\mathcal{H}om(\mathrm{pr}_2^*(G),$$
$$\mathrm{pr}_1^\alpha(F)))\longrightarrow H^\circ(e,\mathcal{O}_*)\quad .$$

Indeed it is enough to consider the composition of the "cup prpduct" with respect to the pairing 7.4.2, and the canonical morphism

$$(7.5.2)\qquad H^\circ(X\times Y,\ K_{X\times Y}) \longrightarrow H^\circ(e,\mathcal{O}_*)$$

induced by the trace morphism

$$\mathbb{R}\,h_*(K_{X\times Y}) \longrightarrow \mathcal{O}_*$$

Definition 7.6. The "Lefschetz Number" Lef(f,u) will be by definition

$$\mathrm{Lef}(f,u) = \mathrm{Lef}(F,F)(c(\Delta),c(f,u))$$

Observation 7.7. The effective calculation of Lef(f,u) is not a trivial problem.

In order to give an example, namely the case of a non-singular curve X with the étale topology, we need some results about discrete valuation rings.

§8. Some results about discrete valuation rings.

Proposition 8.1. Let R be a discrete valuation ring, $\underline{m}$ its maximal ideal, $k = R/\underline{m}$ the residuel field, $f:R \longrightarrow R$ an endomorphism of R and G a finite group of automorphisms of R , which induces the identity on k . We assume there exists an endomorphism

$$\phi:G \longrightarrow G$$

of G such that

$$f(gx) = \phi(g)f(x) \quad \forall g\varepsilon G, \forall x\varepsilon R$$

and that $\nu(f) = v(f(\pi)-\pi) > 1$

If we denote for every $g\varepsilon G$,

$$c^{\phi}(g) = \mathit{Card}\{x\varepsilon G/\phi(x)gx^{-1} = g\}$$

and if we decompose $c^{\phi}(g)$ in the form:

$$c^{\phi}(g) = p^r q \quad , \ (p,q) = 1 \quad , \ p = \mathrm{char}(k) \ ,$$

then $q \mid (\nu(g^{-1}f) - 1)$.

Consequently the rational number

$$\frac{(\nu(g^{-1}f)-1)}{c^{\phi}(g)}$$

is of the form $\frac{\alpha}{p^r}$, $\alpha\varepsilon\mathbf{N}$.

The proof of this proposition uses essentially the following lemma.

Lemma 8.2. Under the same notations as in the proposition 8.1 and under the hypotheses that $\phi = id_G$, if we denote

$$c = \mathit{Card}(\mathrm{Im}(G \longrightarrow \mathrm{Aut}_k(\underline{m}/\underline{m}^2)))$$

We have the following relation of divisibility:

$$c \mid (\nu(f) - 1)$$

Remark 8.3. If A is an algebra over $\mathbf{Z}/_{\ell^{\nu}\mathbf{Z}}$, $(p,\ell) = 1$, then for every element $a\varepsilon A$,

$$\left(\frac{\nu(g^{-1}f)-1}{c^{\phi}(g)}\right) a$$

is a well defined element of A .

Remark 8.4. Grothendieck conjectures that the rational number

$$\frac{\nu(g^{-1}f)-1}{c^{\phi}(g)}$$

is in fact a natural number.

Now suppose we have an A-module M, projective and of finite type over A, which is a G-module and

$$u_M : M \longrightarrow M$$

is an endomorphism of M such that

$$u_M(gx) = \phi(g)u_M(x) \quad \forall g\in G, \ \forall x\in M \ .$$

With this situation we associate an element of A defined by the following formula:

$$(8.5.) \qquad S_A(f,u_M,G) = \sum_{g\in G_{\not\phi}} \left(\frac{\nu(g^{-1}f)-1}{c^{\phi}(g)} \right) Tr_A(u_M g_M^{-1}) \ ,$$

where $G_{\not\phi}$ is the quotient set obtained from G by the equivalence relation:

$$x \sim_{\phi} y \overset{\text{Def}}{\Longleftrightarrow} \exists g \text{ s.t. } y = \phi(g)xg^{-1} \ .$$

Like in the classical theory of zeta function we can introduce the formal series in one variable with coefficients in A:

$$\sum_{n\geq 1} S_A(f^n,u_N^n,G)T^n$$

<u>Conjecture 8.6</u>. (Grothendieck). There exists a formal series

$$\zeta_A(f,u_M,G) = 1+b_1T+b_2T^2+\ldots+b_nT^n+\ldots, \quad b_i\varepsilon A \ ,$$

such that

$$(8.6.1) \qquad T\frac{\zeta_A'(f,u_M,G)}{\zeta_A(f,u_M,G)} = \sum_{n\geq 1} S_A(f^n,u_M^n,G)T^n \ .$$

<u>Proposition 8.7</u>. The conjecture 8.6 is true in the particular case when $G = \{e\}$.

In this case it is enough to find a formal series

$$\zeta_A(f) = 1 + b_1T + b_2T^2 + \ldots b_nT^n + \ldots, \quad b_i\varepsilon \mathbb{Z} \ ,$$

such that

(8.7.1) $$T\frac{\zeta_A'(f)}{\zeta_A(f)} = \nu_1 T + \nu_2 T^2 + \ldots + \nu_r T^r + \ldots$$

where $\nu_i = \nu(f^i)$.

To prove this one uses the relations of Sen (Ann of Math...):

(8.7.2) $$\delta_n = \nu(f^{p^n}) - \nu(f^{p^{n-1}}) \equiv 0 \mod p^n ,$$

and the recursive definition of b_n in function of ν_i :

(8.7.3) $$ib_i = \nu_i + \nu_{i-1}b_1 + \ldots + \nu_1 b_{i-1} , \quad b_1 = \nu_1 .$$

One finds for instance for the first b_i , the formulas:

$$b_i = (\nu_1, i) \quad , i < p$$

$$b_p = (\nu_1, p) + (\nu_1, 0)A_1^1 ,$$

$$b_{p+i} = (\nu_1, p+i) + (\nu_1, i)A_1^1 \quad , i < p$$

$$b_{2p} = (\nu_1, 2p) + (\nu_1, p)A_1^1 + (\nu_1, 0)A_2 ,$$

....

$$b_p^2 = (\nu_1, p^2) + (\nu_1, p^2-p)A_1^1 + (\nu_1, p^2-2p)A_2^1 + \ldots + (\nu_1, p)A_{p-1}^1 + (\nu_1, 0)A_1^2 .$$

....

where

$$(n,i) = \frac{n(n+1)\ldots(n+i-1)}{i!} , \quad A_i^n = \frac{\delta_n(\delta_n+p^n)(\delta_n+2p^n)\ldots(\delta_n+(i-1)p^n)}{p^n.2p^n.3p^n\ldots ip^n}$$

Now let X be a connected normal curve proper over k and

$$f : X \longrightarrow X$$

an endomorphism of X .

We assume that there exists a connected normal curve X' , proper over k , and a finite group G of automorphisms of X' such that X'/G exists and is isomorphic to X .

Let p be the canonical projection:

$$p: X' \longrightarrow X'/G = X \quad .$$

We suppose there exists an endomorphism

$$f': X' \longrightarrow X'$$

of X' and an endomorphism

$$\phi: G \longrightarrow G$$

such that the following conditions are satisfied:

i) $f'(gx') = \phi(g)f'(x)$

ii) the endomorphism f of X is obtained from f' by passing to the quotient.

Under these assumptions we can associate to every fixed point x of f some local numerical invariants.

Let g be an element of G, $X'_x = p^{-1}(x)$ the fiber of p over x and $f'_g = g^{-1}f$. We consider the rational number:

$$(8.8.1) \qquad S_x(f,f',G,g) = \frac{1}{c^{\phi}(g)} \sum_{x' \varepsilon X'^{f'_g}_x} (\nu_{x'}(f'_g) - 1)$$

<u>Proposition 9.1</u>. With the notations of 9.0, the rational number $S_x(f,f',G,g)$ has the form:

$$S_x(f,f',G,g) = \frac{\alpha}{p^r} \, , \ \alpha \varepsilon \mathbb{N} \, , \quad p = \mathrm{char}(k) \quad .$$

<u>Proof</u>. First of all we introduce the endomorphism $\psi: G \longrightarrow G$ defined by:

$$\psi = \mathrm{int}_g \circ \phi \quad , \quad \mathrm{int}_g(x) = g^{-1}xg \quad .$$

It is easy to see that:

i) $\psi(G_{x'}) \subset G_{x'}$, where $G_{x'}$, is the stabilizer of x'.

ii) $c^{\phi}(g) = \mathrm{Card}(G^{\psi})$, $G^{\psi} = \{x \varepsilon G / \psi(x) = x\}$.

iii) G^{ψ} operates on $X'^{f'_g}_x$.

iv) $\nu_{x'}(f'_g) = \nu_{hx'}(f'_g)$ for every $h \in G^{\psi}$.

Consequently we can decompose $X'^{f'_g}_x$ as the sum of its orbits and we obtain using iv):

$$(9.1.1) \quad \sum_{x' \in X'^{f'_g}_x} (\nu_{x'}(f'_g)-1) = \sum_{x' \in X'^{f'_g}_x / G^{\psi}} \frac{(\nu_{x'}(f'_g) - 1)}{\mathrm{Card}(G^{\psi}_{x'})} \mathrm{Card}(G^{\psi}) \text{ , i.e.,}$$

$$(9.1.2) \quad S_x(f,f',G,g) = \sum_{x' \in X'^{f'_g}_x / G^{\psi}} \frac{(\nu_{x'}(f'_g)-1)}{\mathrm{Card}(G^{\psi}_{x'})}$$

and the proposition results from 8.1.

9.2. Let X be a scheme of demension one normal and proper over k , k being an algebraicaly closed field of characteristic $p > 0$, and A an algebra over $\mathbb{Z}/\ell^{\nu}\mathbb{Z}$, ℓ being prime with the p .

We assume that F is a constructible sheaf of A_X-modules (A_X being the constant sheaf over X in étale topology, associated to A) , $f:X \longrightarrow X$ an endomorphism of X with isolated fixed points and $u:f^*(F) \longrightarrow F$ a "lifting" of f .

We suppose:

1) There exists a scheme X' with all conditions in 9.0.

2) There exists a finite set $Y' \subset X'$ stable by G , such that if we introduce the notations:

$p:X' \longrightarrow X'/G = X$, the canonical projection,

$Y = p(Y')$

$U' = X' - Y'$, $U = X - Y$

$p_{U'}:U' \longrightarrow U$, $p_{U'} = p|U'$,

the following conditions hold:

a) $p_{U'}:U' \longrightarrow U$ is a principal covering with structural group G .

b) $p^*(F)|U$ is a constant sheaf generated by an A-module

M , that is $p^*(F)|U' = M_{U'}$

c) There exists a morphism

$$u_{M_{U'}} : (U',M_{U'}) \longrightarrow (U',M_{U'})$$

such that

$$u_{M_{U'}}(gx) = \phi(g)\, u_{M_{U'}}(x') \quad .$$

It is now clear that M has a structure of A[G]-module and has an endomorphism

$$u_M : M \longrightarrow M$$

which satisfies the condition

$$u_M(\phi(g)m) = g u_M(m) \quad .$$

Using (8.5) and (9.1.2) we can introduce the local terms associated with $x \in X^f$.

(9.2.1) $$\alpha_x(f,u,G) = \sum_{g \in G_{f\phi}} S_x(f,f',G,g)\, Tr_A(u_M g_M^{-1})$$

(9.2.2) $$\tau_x(f,u,G) = Tr_A(f_{F_x}) - \alpha_x(f,u,G)$$

Proposition 9.3. (GROTHENDIECK-VERDIER) . With the notations and hypotheses of 9.0, 9.1, 9.2, we have:

(9.3.1) $$Lef(f,u) = \sum_{x \in X^f} \tau_x(f,u,G) \quad .$$

Bibliography

1. Hartshorne, R. Residues and Duality. Lecture Notes in Mathematics, Springer 20(1966).

2. Illusie L. Formule de Lefschetz-Verdier, Séminaire de Geometrie Algébrique, I.H.E.S. 1965/66, Exposé 2.

3. Verdier J.L. The Lefschetz Fixed Point Formula in Etale Cohomology, Proceedings of a Conference in Local Fields, Springer Verlag. 1967, pag. 199-214.

COTANGENT COMPLEX AND DEFORMATIONS OF TORSORS AND GROUP SCHEMES

by Luc Illusie (*)

In this exposé we outline some applications of the cotangent complex to obstruction problems concerning the first order deformations of flat group schemes locally of finite presentation and of torsors under such groups. An enlarged version with detailed proofs will appear in [13]. The results presented here were conjectured by Grothendieck in 1968 and 1969. Those dealing with deformations of flat commutative group schemes play an essential role in recent work of his and W. Messing on Barsotti-Tate groups ([11], [16]).

In § 1 we recall some basic facts about cotangent complex theory. For details the reader is referred to [12]. The main result concerns the obstruction to the first order deformation of a flat ringed topos. Thanks to a general method of deformation of diagrams, which owes much to Deligne's cohomological descent theory (SGA 4 VI), we can apply this result to the problems mentioned above. The diagram we use to handle the deformations of flat commutative group schemes with ring of operators was suggested to us by L. Breen's recent work on the structure of $Ext^i(G_a, G_a)$ ([5]).

1. Review of cotangent complex theory.

1.1. To each map of schemes (more generally, of ringed topoi)

$$f : X \longrightarrow Y$$

is associated a chain complex of flat $\underline{O}_X$-Modules, denoted by

$$L_{X/Y} \quad ,$$

and called the cotangent complex of f (or X over Y), which generalizes in a natural way the complex associated by André [1] and Quillen [18] to

---

(*) part of this research was done while the author was supported by M. I. T.

a map of rings. It is augmented towards $\Omega^1_{X/Y}$, the sheaf of Kähler differentials of f, the augmentation establishing an isomorphism

$$H_o(L_{X/Y}) \xrightarrow{\sim} \Omega^1_{X/Y} \quad .$$

1.2. Suppose f is a morphism of schemes. Then the homology sheaves $H_i(L_{X/Y})$ are quasi-coherent $\underline{O}_X$-Modules. If f is smooth, $\Omega^1_{X/Y}$ is locally free of finite rank and the augmentation $L_{X/Y} \longrightarrow \Omega^1_{X/Y}$ is a quasi-isomorphism (the converse being true when f is locally of finite presentation). If f is a locally complete intersection map in the sense of Berthelot (SGA 6 VIII 1), then, in the terminology of (SGA 6 I 4.8), $L_{X/Y}$ is of perfect amplitude $\subset [-1,0]$, which means that $L_{X/Y}$ is locally isomorphic, in the derived category D(X), to a complex of locally free sheaves of finite rank concentrated in degrees -1 and 0 .(The converse (due to Quillen) is true provided that Y is locally noetherian and f locally of finite type.) Finally, suppose f admits a factorization

$$X \xrightarrow{i} X' \xrightarrow{f'} Y \quad ,$$

where f' is formally smooth and i is a closed immersion defined by an Ideal I . Then, in D(X) there is a canonical isomorphism

$$t_{[-1}(L_{X/Y}) \simeq (0 \rightarrow I/I^2 \xrightarrow{d} i^*(\Omega^1_{X'/Y}) \rightarrow 0) \quad ,$$

where $i^*(\Omega^1_{X'/Y})$ is placed in degree 0, d is induced by the universal derivative $d_{X'/Y} : \underline{O}_{X'} \longrightarrow \Omega^1_{X'/Y}$, and $t_{[n}(L)$, for a complex L, denotes the complex deduced from L by killing $H^i(L)$ for $i < n$, namely $(0 \rightarrow L^n/B^n \rightarrow L^{n+1} \rightarrow L^{n+2} \rightarrow \ldots)$.

1.3. $L_{X/Y}$ depends functorially on f in the same way as $\Omega^1_{X/Y}$. This means that each (essentially) commutative square of ringed topoi

(1.3.1)
$$\begin{array}{ccc} X & \xleftarrow{g} & X' \\ {\scriptstyle f}\downarrow & & \downarrow \\ Y & \longleftarrow & Y' \end{array}$$

gives rise to a map of complexes

(1.3.2)
$$g^*L_{X/Y} \longrightarrow L_{X'/Y'} \quad ,$$

these maps satisfying a certain coherence condition relative to the composition of squares. Note that $g^*L_{X/Y} = Lg^*L_{X/Y}$ in $D(X)$ since the components of $L_{X/Y}$ are flat.

__Proposition__ 1.3.3 (base change). __Suppose__ (1.3.1) __is defined by a cartesian square of schemes such that__ $\underline{Tor}_i^{Y}(\underline{O}_X, \underline{O}_{Y'}) = 0$ for $i > 0$ (__which is the case for example if__ X __or__ Y' __is flat over__ Y). __Then__ (1.3.2) __is a quasi-isomorphism.__

1.4. Let

$$(1.4.1) \qquad X \xrightarrow{f} Y \longrightarrow Z$$

be maps of ringed topoi. Then there is a canonical, exact triangle in $D(X)$

$$(1.4.2) \qquad \begin{array}{ccc} & L_{X/Y} & \\ \swarrow & & \nwarrow \\ f^*L_{Y/Z} & \longrightarrow & L_{X/Z} \end{array} ,$$

where the maps of degree 0 are those defined by functoriality of the cotangent complex. It is called the __transitivity triangle__. It depends functorially on (1.4.1). The map of degree 1 in (1.4.2) is sometimes denoted by $K(X/Y/Z)$, and called the __Kodaira-Spencer map__ (or __class__). When (1.4.1) is defined by smooth morphisms of schemes, $K(X/Y/Z)$ coïncides with the usual class in $H^1(X, T_{X/Y} \otimes f^*\Omega^1_{Y/Z})$, where $T_{X/Y}$ is the tangent sheaf of f (dual of $\Omega^1_{X/Y}$).

1.5. Let $f : X \rightarrow Y$ be a map of ringed topoi, and let M be an $\underline{O}_X$-Module. By a Y-__extension of__ X __by__ M we mean a factorization

$$\begin{array}{ccc} X & \xrightarrow{i} & X' \\ {\scriptstyle f}\downarrow & \swarrow & \\ Y & & \end{array}$$

where i is an equivalence on the underlying topoi and on the rings induces a surjective map $\underline{O}_{X'} \rightarrow \underline{O}_X$ whose kernel is of square zero and isomorphic to M as an $\underline{O}_X$-Module. Maps of Y-extensions are defined in the obvious way. Note that if f is defined by a map of schemes and M is quasi-coherent, then the above factorization comes from a factorization in the category of schemes.

The interest of the cotangent complex in the theory of deformations comes from :

Theorem 1.5.1. *There exists a canonical, functorial isomorphism between the set of isomorphism classes of* Y-*extensions of* X *by* M, *equipped with the group structure defined by the usual addition law on extensions, and the group* $\mathrm{Ext}^1(L_{X/Y},M)$. *Moreover, the group of automorphisms of any fixed* Y-*extension* X' *of* X *by* M *is canonically isomorphic to* $\mathrm{Ext}^0(L_{X/Y},M)$.

(The last part is, of course, a trivial consequence of the isomorphism $H_0(L_{X/Y}) \xrightarrow{\sim} \Omega^1_{X/Y}$ (1.1)).

1.6. Consider a commutative diagram of ringed topoi

$$(1.6.1)\qquad \begin{array}{ccc} X & \xrightarrow{i} & X' \\ {\scriptstyle f}\downarrow & & \downarrow{\scriptstyle f'} \\ Y & \xrightarrow{j} & Y' \\ \downarrow & \swarrow & \\ S & & \end{array}$$

where j is an S-extension of Y by some $\underline{O}_Y$-Module J and i is a Y'-extension of X by some $\underline{O}_X$-Module I . Such a diagram defines a map of $\underline{O}_X$-Modules

$$(*)\qquad f^*J \longrightarrow I \qquad .$$

We shall call f' a *deformation of* f *over* Y' if (*) is an isomorphism. If f is flat, then any deformation f' is automatically flat, as a result of the well-known flatness criterion. The key result in deformation theory is the following, which is a formal consequence of (1.4) and (1.5.1) :

Theorem 1.7. *Let*

$$X \xrightarrow{f} Y \longrightarrow S$$

be maps of ringed topoi, and let $j : Y \longrightarrow Y'$ *be an* S-*extension of* Y *by an* $\underline{O}_Y$-*Module* J . *Then* :

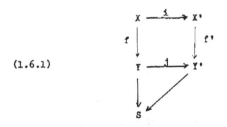

(i) <u>There exists an obstruction</u>

$$\omega(f,j) \in \mathrm{Ext}^2(L_{X/Y}, f^* J)$$

<u>whose vanishing is necessary and sufficient for the existence of a deformation</u> f' <u>of</u> f <u>over</u> Y' .

(ii) <u>When</u> $\omega(f,j) = 0$, <u>the set of isomorphism classes of deformations</u> f' <u>is an affine space under</u> $\mathrm{Ext}^1(L_{X/Y}, f^* J)$ <u>and the group of automorphisms of a fixed deformation is canonically isomorphic to</u> $\mathrm{Ext}^0(L_{X/Y}, f^* J)$.

(iii) <u>The obstruction</u> $\omega(f,j)$ <u>can be written as a Yoneda cup-product</u>

$$\omega(f,j) = (f^* e(j)) K(X/Y/S)$$

<u>where</u> $K(X/Y/S) \in \mathrm{Ext}^1(L_{X/Y}, f^* L_{Y/S})$ <u>is the Kodaira-Spencer class</u> (1.4) <u>and</u> $e(j) \in \mathrm{Ext}^1(L_{Y/S}, J)$ <u>is the class defined by</u> j (1.5.1).

2. <u>Equivariant deformations</u>.

2.1. Fix a scheme S and a group scheme G over S . Let X, Y be G-schemes, and let $f : X \longrightarrow Y$ be a G-equivariant map. We shall assume G or f to be flat. To avoid technical complications, we shall also assume f to be a complete intersection map in the sense of Berthelot (SGA 6 VIII 1) (recall this implies (1.2) that $L_{X/Y}$ is a perfect complex). Taking into account the action of G, and using the base change property (1.3.3), we can define a complex of G-$\underline{O}_X$-Modules, or more precisely an object

(2.1.1) $$L^G_{X/Y} \in \mathrm{ob}\, D^b(BG_{/X}) \quad ,$$

unique up to unique isomorphism, in such a way that the underlying complex of $\underline{O}_X$-Modules is canonically isomorphic to $L_{X/Y}$ in $D(X)$, and $H_0(L^G_{X/Y}) \xrightarrow{\sim} \Omega^1_{X/Y}$ as a G-$\underline{O}_X$-Module ([1]). Here BG means the classifying topos (SGA 4 IV 2.4) of G considered as a sheaf of groups for the fpqc topology on the category of

---

([1]) These conditions do not characterize $L^G_{X/Y}$!

all schemes over S, and $BG_{/X}$ is the topos of objects of BG over X ; BG (hence $BG_{/X}$) is equipped with the canonical ring defined by the structural rings of schemes over S. When X admits an equivariant closed embedding $i : X \longrightarrow X'$ into a smooth G-scheme X' over Y, $L^G_{X/Y}$ can be taken to be the complex of G-$\underline{O}_X$-Modules (cf. (1.2))

$$0 \longrightarrow I/I^2 \xrightarrow{\ d\ } i^*(\Omega^1_{X'/Y}) \longrightarrow 0$$

where I is the Ideal of i . In the general case, the definition of (2.1.1) is a little involved ; some indications are given below. Note that if X, Y are trivial G-schemes, then we have $L^G_{X/Y} \xrightarrow{\sim} L_{X/Y}$ where $L_{X/Y}$ is viewed as a complex of trivial G-$\underline{O}_X$-Modules.

The equivariant cotangent complex $L^G_{X/Y}$ enjoys functorial properties analogous to those satisfied by $L_{X/Y}$. In particular, a composition of G-maps gives rise to an "equivariant transitivity triangle", hence to an "equivariant Kodaira-Spencer class". Details will be omitted.

2.2. Consider a commutative diagram of G-schemes over S of the form (1.6.1) where all maps are G-maps and i (resp. j) is a closed immersion defined by an Ideal I (resp. J) of square zero. So I (resp. J) is a G-$\underline{O}_X$- (resp. G-$\underline{O}_Y$-) Module and the canonical map $f^*J \rightarrow I$ is a G-map. We shall call f' an <u>equivariant deformation of</u> f <u>over</u> Y' if the underlying map of schemes is a deformation of f in the sense of (1.6), or equivalently if the canonical map $f^*J \rightarrow I$ is a G-isomorphism.

The following result is a consequence of (1.7). A sketch of proof will be given below.

<u>Proposition</u> 2.3. <u>Suppose</u> G <u>and</u> $f : X \longrightarrow Y$ <u>satisfy the hypotheses of</u> (2.1) . <u>Fix a closed equivariant embedding</u> $j : Y \rightarrow Y'$ <u>of</u> G-<u>schemes over</u> S, <u>the Ideal</u> J <u>of</u> j <u>being of square zero</u>. <u>Then there is an obstruction</u>

$$\omega(G,f,j) \in \operatorname{Ext}^2_G(L^G_{X/Y}, f^*J)$$

<u>whose vanishing is necessary and sufficient for the existence of an</u>

equivariant deformation f' of f over Y'. When $\omega(G,f,j) = 0$, the set of isomorphism classes of equivariant deformations f' is an affine space under $\mathrm{Ext}^1_G(L^G_{X/Y}, f^*J)$, and the group of automorphisms of a fixed f' is canonically isomorphic to $\mathrm{Ext}^0_G(L^G_{X/Y}, f^*J)$. (Here the notation $\mathrm{Ext}^i_G(L,M)$ for $L, M \in \mathrm{ob}\, D(BG_{/X})$ is short for $\mathrm{Ext}^i(BG_{/X}; L,M)$.)

It is also possible to write $\omega(G,f,j)$ as a cup-product with an equivariant Kodaira-Spencer class. A precise statement will be given in the particular case of torsors, which we are now going to examine.

2.4. From now on we shall assume G to be flat and locally of finite presentation. By the well-known theorem giving the local structure of algebraic groups this implies that $G \to S$ is a (locally) complete intersection map (in the sense of Berthelot ([1])). Let $f : X \to Y$ be a G-map of G-schemes over S, the action of G on Y being trivial. Denote by G_Y the group scheme over Y induced by G (i.e. $G_Y = G \times_S Y$). Recall that X is said to be a principal homogeneous space (or torsor ([2])) under G_Y if the following conditions are satisfied :

(i) the map $G_Y \times_Y X \longrightarrow X \times_Y X$, $(g,x) \longmapsto (gx,x)$ is an isomorphism

(ii) $f : X \to Y$ is faithfully flat and quasi-compact.

These also amount to saying that after some fpqc base change $Y' \to Y$ (for example, f) X becomes isomorphic to G_Y, acting on itself by left multiplication. They imply f is a complete intersection (because the latter property is local for the fpqc topology (SGA 6 VIII 1.6)). So, by (2.1) the equivariant cotangent complex $L^G_{X/Y}$ is defined. Denote by $f^G : BG_{/X} \to Y$ the canonical map (f^G_* is "taking the global sections invariant under G"). It can be shown by descent that $Rf^G_*(L^G_{X/Y})$ is a perfect complex, of perfect amplitude $\subset [-1,0]$ (SGA 6 I 4.8), and that the adjunction map

([1]) which here is also the sense of (EGA IV 19.3.6) because of (SGA 6 VIII 1.4).

([2]) from the French "torseur".

$$Lf^{G*}Rf_*^G(L_{X/Y}^G) \longrightarrow L_{X/Y}^G$$

(hence, too, the adjunction map $Lf^* \; Rf_*^G(L_{X/Y}^G) \longrightarrow L_{X/Y}$) is an isomorphism.

<u>Definition</u> 2.5. <u>Let</u> Y <u>be a scheme over</u> S <u>and let</u> $f : X \to Y$ <u>be a torsor under</u> G_Y . <u>The complex</u> $Rf_*^G(L_{X/Y}^G) \in \text{ob } D^b(Y)$ <u>will be denoted by</u> $\ell_{X/Y}$ <u>and called the co-Lie complex of</u> X <u>over</u> Y.

2.5.1. The formation of the co-Lie complex commutes with any base change $Y' \to Y$. If X is trivial over Y, i.e. f admits a section $s : Y \to X$, then we have

$$\ell_{X/Y} \simeq Ls^*(L_{X/Y}) \quad .$$

In particular, take Y = S and X to be G acting on itself by left multiplication. Then we have

$$\ell_{G/S} \simeq Le^*(L_{G/S})$$

where $e : S \to G$ is the unit section. The complex $\ell_{G/S}$, often denoted simply ℓ_G, is called the co-Lie complex of G, and will be discussed later. It was first introduced by Mazur-Roberts [15] for a finite G. When G is smooth, it coincides with the sheaf of invariant differential forms ω_G, dual to the Lie algebra of G .

<u>Theorem</u> 2.6. <u>In the situation of</u> (2.5), <u>let</u> $j : Y \hookrightarrow Y'$ <u>be an</u> S-<u>extension of</u> Y <u>by a quasi-coherent</u> $\underline{O}_Y$-<u>Module</u> J . <u>Then there exists an obstruction</u>

$$\omega(G,f,j) \in H^2(Y, \ell_{X/Y}^{\vee} \overset{L}{\otimes} J) \qquad (^1)$$

<u>whose vanishing is necessary and sufficient for the existence of a torsor</u> $f' : X' \to Y'$ <u>under</u> $G_{Y'}$, <u>such that</u> $X' \times_{Y'} Y$ <u>be isomorphic to</u> X (<u>as a torsor under</u> G_Y). <u>When</u> $\omega(G,f,j) = 0$, <u>the set of isomorphism classes of such torsors</u> X' <u>is an affine space under</u> $H^1(Y, \ell_{X/Y}^{\vee} \overset{L}{\otimes} J)$, <u>and the group of automorphisms of a solution is canonically isomorphic to</u> $H^0(Y, \ell_{X/Y}^{\vee} \overset{L}{\otimes} J)$.

---

(1) For $L \in \text{ob } D(Y)$ we set $L^{\vee} = R\underline{Hom}(L, \underline{O}_Y)$.

In effect, it is easily seen that a torsor X' under $G_{Y'}$ inducing X on Y is the same as an equivariant deformation of f over Y' (2.2). Therefore the theorem follows from (2.3), since by descent we have

$$\begin{aligned}\mathrm{Ext}^i_G(L^G_{X/Y}, f^*J) &\simeq \mathrm{Ext}^i(\chi_{X/Y}, J)\\ &\simeq H^i(Y, \chi^\vee_{X/Y} \overset{L}{\otimes} J) \qquad (\chi_{X/Y} \text{ being perfect}).\end{aligned}$$

__Remark__ 2.6.1. Suppose G is smooth. Then $\chi^\vee_{X/Y}$ is nothing but the Lie Algebra t_{G_Y} of G_Y twisted by the torsor X via the adjoint action of G_Y :

$$\chi^\vee_{X/Y} \simeq t_{X/Y} \overset{\mathrm{dfn}}{=} X \overset{G_Y}{\times} t_{G_Y} \qquad ,$$

and (2.6) is easy to prove directly. Observe that equivariant deformations of X locally exist and that any two deformations are always locally isomorphic, the sheaf of automorphisms of a given deformation being identified with $t_{X/Y}$. Hence the obstruction in H^2 is obtained by a classical cocycle calculation best expressed in Giraud's language of "gerbes" [9].

2.7. It is possible by descent to construct from the Kodaira-Spencer class (1.4) a canonical class, called the __Atiyah class__ of X,

(2.7.1) $\quad at(X/Y/S) \in \mathrm{Ext}^1(\chi_{X/Y}, tL_{Y/S})$

where $tL_{Y/S}$ stands for the pro-object of truncated complexes $``\varprojlim_n" \, t_{[n} L_{Y/S}$ (1). As in (1.7), the obstruction $\omega(G,f,j)$ of (2.6) can be written as a cup-product

(2.7.2) $\quad \omega(G,f,j) = e(j)at(X/Y/S)$

where $e(j) \in \mathrm{Ext}^1(L_{Y/S}, J) = \mathrm{Ext}^1(tL_{Y/S}, J)$ is the class of the Y-extension j . When both G and Y are smooth, at(X/Y/S) coincides with the class in $H^1(Y, t_{X/Y} \otimes \Omega^1_{Y/S})$ constructed by Atiyah in [3] .

2.8. __Sketch of proof of__ (2.3). a) Recall that to any G-scheme X over S is associated in a functorial way a simplicial scheme over S, called its __nerve__ :

$$\mathrm{Ner}(G,X) = (\ldots G^n \times X \rightrightarrows \quad \ldots G \times X \rightrightarrows X)$$

---

(1) for the notation $t_{[n}$, see (1.2).

(products are taken over S, faces and degeneracies are given by the standard formulas :

$$d_0(g_1,\ldots,g_n,x) = (g_2,\ldots,g_n,g_1x)$$
$$d_1(g_1,\ldots,g_n,x) = (g_1g_2,g_3,\ldots,g_n,x) \quad ,$$

etc.). So, in the situation of (2.3), we have a commutative diagram of simplicial schemes over S :

$$\begin{array}{ccc} \mathrm{Ner}(G,X) & & \\ \downarrow & & \\ \mathrm{Ner}(G,Y) & \longrightarrow & \mathrm{Ner}(G,Y') \\ \downarrow & \swarrow & \\ \mathrm{Ner}(G) & & \end{array} \qquad (2.8.1)$$

where Ner(G) is short for Ner(G,S). Thanks to the hypotheses and the flatness criterion, it is easily checked that an equivariant deformation of X over Y' is the same as a deformation of Ner(G,X) over Ner(G,Y') as a simplicial scheme, i.e. a commutative square of simplicial schemes

$$\begin{array}{ccc} \mathrm{Ner}(G,X) & \longrightarrow & Z \\ \downarrow & & \downarrow \\ \mathrm{Ner}(G,Y) & \longrightarrow & \mathrm{Ner}(G,Y') \end{array}$$

such that, for each $n \in \mathbb{N}$, $Z_n \longrightarrow G^n \times Y'$ is a deformation of $G^n \times X \longrightarrow G^n \times Y$ (as a map of schemes).

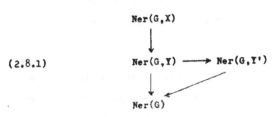

b) Recall that any diagram D ($i \mapsto D_i$) of ringed spaces defines (SGA 4 VI) a ringed topos still denoted by D and called the <u>total topos</u> of D . A Module on D consists of a family of Modules E_i on the D_i together with transition maps $f^*E_j \longrightarrow E_i$ satisfying certain compatibility conditions. So (2.8.1) defines a diagram of ringed topoi, in which the triangle is a Ner(G)-extension of Ner(G,Y) by an Idéal of square zero, which will still be denoted by J for simplicity. Moreover, a deformation Z as above is the same as a deformation of Ner(G,X) $\longrightarrow$ Ner(G,Y) over Ner(G,Y') as a map of ringed topoi. Therefore we can apply (1.7) and the problem boils down to identifying the groups $\mathrm{Ext}^i(L_{\mathrm{Ner}(G,X)/\mathrm{Ner}(G,Y)}, f^* J)$ with the groups $\mathrm{Ext}^i_G(L^G_{X/Y}, f^* J)$ of (2.3).

c) First of all we have to define $L^G_{X/Y}$. Let us indicate very briefly how to do this. By an analogue of the nerve construction, we get a map of ringed topoi $\mathrm{Ner}(G,X) \longrightarrow BG_{/X}$, which, by descent, induces a fully faithful functor

$$(*) \qquad D^b(BG_{/X})_{qcoh} \longrightarrow D^b(\mathrm{Ner}(G,X))$$

where $D^b(BG_{/X})_{qcoh}$ is the full sub-category of $D(BG_{/X})$ defined by those complexes whose cohomology is bounded and quasi-coherent. Now, thanks to the hypotheses and the base changed property (1.3.3), it can be shown that $L_{\mathrm{Ner}(G,X)/\mathrm{Ner}(G,Y)}$ is isomorphic to the image under (*) of an object $L^G_{X/Y}$ of $D^b(BG_{/X})_{qcoh}$ (unique up to unique isomorphism). The identification desired at the end of b) follows from the fact that (*) is fully faithful. This completes the proof of (2.3). The Atiyah class (2.7.1) and the formula (2.7.2) are easily deduced from the Kodaira-Spencer class of the vertical composition in (2.8.1) and from (1.7 (iii)).

Remark 2.9. There is an alternate approach to the results of this section, which is based on Deligne's theory of Picard stacks (SGA 4 XVIII). It also yields interesting refinements, for example the following, which will be appreciated by the specialist : in the situation of (2.6), assume $\omega(G,f,j) = 0$: then the Picard stack of equivariant deformations of X over Y' is represented by the complex $j_*(\mathcal{X}^\vee_{X/Y} \overset{L}{\otimes} J)[1]$.

3. Deformations of non-commutative, flat group schemes.

In this section, we fix a scheme S and a flat, locally of finite presentation group scheme G over S.

3.1. Using the action of G on itself by left multiplication, we have defined in (2.5.1) the co-Lie complex of G, $\mathcal{X}_G \in \mathrm{ob}\, D(S)$, from which we can reconstruct $L_{G/S}$ by means of the canonical isomorphism (2.4.1)

$$Lf^*(\mathcal{X}_G) \xrightarrow{\sim} L_{G/S}$$

where $f : G \to S$ is the projection. We could as well have used the action of G on itself by right multiplication to define analogously an object ℓ_G' of D(S), but ℓ_G and ℓ_G' would have been canonically isomorphic since both have to be canonically isomorphic to $Le^*(L_{G/S})$ where $e : S \to G$ is the unit section. Yet, if we denote by G^o the opposite group and let $G \times G^o$ (1) act on G by $(g,h)x = gxh$, we can define a finer object than ℓ_G, namely

(3.1.1) $\quad \underline{\ell}_G \overset{dfn}{=} Rf_*^{G^o} (L_{G/S}^{G\times G^o}) \in \text{ob } D(BG)$

where $L_{G/S}^{G\times G^o} \in \text{ob } D(B(G\times G^o)_{/G})$ is the equivariant cotangent complex of the $G\times G^o$-scheme G (2.1.1) and $f^{G^o} : B(G\times G^o)_{/G} \longrightarrow BG$ is the canonical map defined by f ($f_*^{G^o}$ is "taking the sheaf of global sections invariant under G^o"). It follows easily from the definition that the object of D(S) deduced from $\underline{\ell}_G$ by forgetting the action of G is canonically isomorphic to ℓ_G. When G is smooth over S, $\underline{\ell}_G$ is nothing but ω_G, the sheaf of right invariant differential forms of degree 1, equipped with the adjoint action of G.

From $\underline{\ell}_G$ we can reconstruct the co-Lie complex of any G-torsor. In effect, let Y be a scheme over S and X be a torsor on Y under G_Y. By the classifying property of BG, X defines a map $p : Y \to BG$ such that $p^*(PG) \simeq X$ where PG is the universal torsor on BG. Then we have

(3.1.2) $\quad \ell_{X/Y} \simeq p^*\underline{\ell}_G$,

which generalizes the formula of (2.6.1).

3.2. Suppose now S is a T-scheme and we are given a T-extension S' of S by a quasi-coherent $\underline{O}_S$-Module I, $i : S \to S'$. We want to pinpoint the obstruction to the existence of a deformation of G over S', by which we mean a flat group scheme G' over S' together with an isomorphism $G' \times_{S'} S \xrightarrow{\sim} G$ (note that if G is commutative, G' need not be commutative). Before stating the main result, we need a notation.

---

(1) unless otherwise stated, products are taken over S.

3.3. Let $u : X \longrightarrow Y$ be a map of ringed topoi, which for simplicity we shall assume to be flat. Denote by Ab the category of abelian groups. In ([12] III 4.10) we define an exact functor

$$R\Gamma(Y/X,-) : D^{+}(Y) \longrightarrow D^{+}(Ab)$$

with the property that, for $E \in \text{ob } D^{+}(Y)$, there exists a canonical, functorial, exact triangle

$$(3.3.1) \qquad \begin{array}{ccc} & R\Gamma(X,u^{*}E) & \\ \swarrow & & \nwarrow {\scriptstyle *} \\ R\Gamma(Y/X,E) & \longrightarrow & R\Gamma(Y,E) \end{array}$$

where $*$ is the restriction map. The cohomology groups

$$H^{i}(Y/X,E) \overset{\text{dfn}}{=} H^{i}R\Gamma(Y/X,E)$$

are called the relative cohomology groups of E mod. Y . They are related to the absolute cohomology groups by the exact sequence of cohomology of (3.3.1) :

$$\ldots \to H^{i}(Y/X,E) \longrightarrow H^{i}(Y,E) \longrightarrow H^{i}(X,u^{*}E) \longrightarrow H^{i+1}(Y/X,E) \longrightarrow \ldots \quad .$$

We can now state :

Theorem 3.4. *In the situation of* (3.2), *there exists an obstruction*

$$\omega(G,i) \in H^{3}(BG/S,\underline{\underline{\ell}}_G^{\vee} \overset{L}{\otimes} I) \qquad (1)$$

whose vanishing is necessary and sufficient for the existence of a deformation of G *into a flat group scheme* G' *over* S'. *When* $\omega(G,i) = 0$, *the set of isomorphism classes of deformations* G' *is an affine space under* $H^{2}(BG/S,\underline{\underline{\ell}}_G^{\vee} \overset{L}{\otimes} I)$, *and the group of automorphisms of a given deformation is canonically isomorphic to* $H^{1}(BG/S,\underline{\underline{\ell}}_G^{\vee} \overset{L}{\otimes} I)$. *Moreover, there exists a canonical class, depending only on* $G \to S \to T$,

$$c(G/S/T) \in H^{2}(BG/S,\underline{\underline{\ell}}_G^{\vee} \overset{L}{\otimes} tL_{S/T}) \quad , (2)$$

whose cup-product with the class $e(i) \in \text{Ext}^{1}(L_{S/T},I)$ *defined by the*

---

(1) For $L \in \text{ob } D(BG)$, $L^{\vee} \overset{\text{dfn}}{=} R\underline{\text{Hom}}(L,\underline{O})$. The relative cohomology groups are taken with respect to the unit section map $S \to BG$.

(2) As in (2.7), $tL_{S/T}$ stands for the pro-object "$\underleftarrow{\lim}$" $t_{[n}L_{S/T}$.

T-<u>extension</u> i <u>yields the obstruction</u> $\omega(G,i)$:

$$\omega(G,i) = e(i)c(G/S/T) \quad .$$

3.5. The above result improves those of Demazure in (SGA 3 III) where, aside from the fact that only the smooth case was discussed, the problem of deforming G as a group scheme was not considered as a whole, but rather broken down into successive steps : (i) deform G as a scheme (ii) deform the multiplication $G \times G \to G$ (iii) render it associative. As Grothendieck pointed out, the partial obstructions encountered in (loc. cit.) turn out to be the images of $\omega(G,i)$ into the successive quotients of the filtration of $H^3(BG/S, \underline{\ell}_G^{\vee} \overset{L}{\otimes} I)$ arising from the "Moore spectral sequence" :

$$E_1^{pq} = \begin{cases} H^q(G^p, \ell_G^{\vee} \overset{L}{\otimes} I) & \text{if } p \geqslant 1 \\ 0 & \text{if } p \leqslant 0 \end{cases} \Longrightarrow H^*(BG/S, \underline{\ell}_G^{\vee} \overset{L}{\otimes} I) \quad .$$

A complete discussion will be found in [13].

3.6. <u>Sketch of proof of</u> (3.4). a) Let Y be a scheme. The functor $G \mapsto \mathrm{Ner}(G)$ (2.8 a)) from the category of group schemes over Y to the category of simplicial schemes over Y is fully faithful, and it is easy to see that its essential image consists of exactly those simplicial schemes X over Y which satisfy the following exactness conditions :

(i) $X_o \to Y$ is an isomorphism ;

(ii) for $n \geqslant 2$, the canonical map

$$X_n \to X_1 \times_{X_o} \cdots \times_{X_o} X_1 \qquad \text{(n factors)} \quad ,$$

obtained by writing the interval $[0,n]$ as an amalgamated sum, is an isomorphism ;

(iii) the squares

$$\begin{array}{ccc} X_2 & \xrightarrow{d_1} & X_1 \\ {\scriptstyle d_2}\downarrow & & \downarrow{\scriptstyle d_1} \\ X_1 & \xrightarrow{d_1} & X_o \end{array} \qquad \text{and} \qquad \begin{array}{ccc} X_2 & \xrightarrow{d_1} & X_1 \\ \downarrow{\scriptstyle d_o} & & \downarrow{\scriptstyle d_o} \\ X_1 & \xrightarrow{d_o} & X_o \end{array}$$

are cartesian. ((ii) expresses the fact that X comes from a category object,

(i) + (ii) that X comes from a monoïd, (ii) + (iii) that X comes from a groupoïd.)

b) Using a) and the flatness criterion, we see that a deformation of G into a flat group scheme over S' is essentially the same as a deformation of Ner(G) into a flat simplicial scheme over S' or, equivalently, a deformation over S' of the corresponding ringed topos (2.8 b)). Therefore we can again apply (1.7), and what remains is to identify the groups $\mathrm{Ext}^i(L_{\mathrm{Ner}(G)/S}, f^* I)$ for $0 \leqslant i \leqslant 2$. Now, by looking at the exact triangle of the composition $\mathrm{Ner}(G,G) \rightarrow \mathrm{Ner}(G) \rightarrow S$, it is not too hard to prove that there is a canonical isomorphism

$$\mathrm{Ext}^*(L_{\mathrm{Ner}(G)/S}, f^* I) \simeq H^{*+1}(BG/S, \underline{\ell}_G^\vee \overset{L}{\otimes} I)$$

(actually I could be replaced by any complex with bounded, quasi-coherent cohomology), and that concludes the proof.

4. Interlude on Lie and co-Lie complexes.

The group schemes considered in this section are assumed to be flat and locally of finite presentation over a fixed scheme S.

4.1. Notations. Let G be a group scheme over S. Recall that the co-Lie complex ℓ_G is of perfect amplitude $\subset [-1,0]$, hence has only two interesting cohomology sheaves, namely

$$\omega_G = H^0(\ell_G) \quad , \qquad n_G = H^{-1}(\ell_G) \quad .$$

The dual of ℓ_G, i.e. $\ell_G^\vee = R\underline{\mathrm{Hom}}(\ell_G, \underline{O}_S)$, is called the Lie complex of G. It is perfect, of perfect amplitude $\subset [0,1]$. Its two possibly non zero cohomology sheaves will be denoted by

$$t_G = H^0(\ell_G^\vee) \quad , \qquad \nu_G = H^1(\ell_G^\vee) \quad .$$

Note we have

$$t_G = (\omega_G)^\vee \quad , \qquad n_G = (\nu_G)^\vee$$

(where $-^\vee = \underline{\mathrm{Hom}}(-, \underline{O}_S)$). Therefore the basic invariants are ω_G, ν_G (and the class in $\mathrm{Ext}^2(\omega_G, n_G)$ (resp. $\mathrm{Ext}^2(\nu_G, t_G)$) defined by ℓ_G (resp. $\ell_G^\vee$)).

We shall first give two general methods for computing ℓ_G.

4.2. Suppose we are given a closed embedding of G into a smooth group scheme over S, $i : G \longrightarrow G'$, and denote by I the Ideal of i . Then, by (1.2) we have

$$L_{G/S} \simeq (0 \to I/I^2 \xrightarrow{d} i^* \Omega^1_{G'/S} \to 0) \quad ,$$

hence, by (2.5.1) :

$$\ell_G \simeq Le^*(L_{G/S}) \simeq (0 \to e^*(I/I^2) \longrightarrow \omega_{G'} \to 0)$$

where $e : S \to G$ is the unit section. Moreover, as i is a homomorphism of groups, G acts naturally on the right hand side, and the complex of G-$\underline{O}_S$-Modules thus defined represents $\underline{\ell}_G$ in D(BG).

4.3. Suppose now we have an exact sequence of group schemes over S

(4.3.1) $\quad 1 \to G \longrightarrow G' \longrightarrow G'' \to 1$

("exact" being taken with respect to the fpqc topology). Then, as Mazur-Roberts observed in [15], there is defined a canonical exact triangle in D(S) :

(4.3.2)
$$\begin{array}{ccc} & \ell_G & \\ \swarrow & & \nwarrow \\ \ell_{G''} & \longrightarrow & \ell_{G'} \end{array}$$

where the horizontal map is induced by $G' \longrightarrow G''$ by functoriality of the co-Lie complex. It is indeed an immediate consequence of (1.3.3) and (1.4.2). Particularly interesting is (4.3.2) when G', G" are smooth. We then obtain :

(4.3.3) $\quad \ell_G \simeq (0 \to \omega_{G''} \longrightarrow \omega_{G'} \to 0) \quad .$

<u>Examples</u> 4.3.4. a) Take $G = (\mu_n)_S$. We have the exact sequence :

$$0 \longrightarrow (\mu_n)_S \longrightarrow (G_m)_S \xrightarrow{n} (G_m)_S \longrightarrow 0 \quad .$$

Hence (4.3.3) yields :

$$\ell_{\mu_{nS}} \simeq (0 \to \underline{O}_S \xrightarrow{n} \underline{O}_S \to 0) \quad .$$

In particular, we have $\ell_{\mu_{nS}} = 0$, i.e. μ_{nS} is étale, if and only if n is invertible on S, which is of course well known.

b) Suppose $p1_S = 0$, where p is a prime number. The group scheme α_p on S is defined by the exact sequence

$$0 \longrightarrow \alpha_p \longrightarrow (G_a)_S \xrightarrow{F} (G_a)_S \longrightarrow 0$$

where F is the Frobenius map $x \mapsto x^p$. Hence (4.3.3) gives

$$\mathcal{X}_{\alpha_p} \simeq (0 \to \underline{O}_S \xrightarrow{0} \underline{O}_S \to 0) \quad .$$

c) Suppose G is a finite, locally free, commutative group scheme over S. Let A be the bi-algebra defining G, and $A^{\vee} = \underline{\mathrm{Hom}}(A, \underline{O}_S)$ denote the dual of A, which is also the bi-algebra of the Cartier dual G^{*} . It is well known (see for instance ([16] II 3.2.4)) that there is a canonical, functorial, closed embedding of G into $W(A^{\vee})^{+} = G'$ ([1]) , hence we have a canonical, functorial exact sequence $0 \to G \to G' \to G'' \to 0$ with G' and G" smooth, which by (4.3.3) provides a canonical, functorial way of calculating $\mathcal{X}_G$.

We shall now discuss some general properties of the Lie and co-Lie complexes in the commutative case. First of all, we have the following result, which generalizes the fact that the adjoint action of a commutative group on its Lie algebra is trivial :

<u>Proposition</u> 4.4. <u>Let</u> G <u>be a commutative group scheme over</u> S <u>and</u> $p : BG \longrightarrow S$ <u>denote the canonical projection</u>. <u>There is a canonical</u>, <u>functorial isomorphism</u> <u>in</u> D(BG) :

$$\underline{\mathcal{X}}_G \simeq Lp^{*}(\mathcal{X}_G) \quad .$$

<u>Proof</u>. See [13]. The basic observation is that the multiplication $m : G \times G \longrightarrow G$, being a group homomorphism, induces a map $B(G \times G) \longrightarrow BG$, for which the inverse image of G as a G-object by left multiplication is G as a (G x G)-object by left and right multiplication.

4.5. Let Y be a scheme over S, and $j : Y \to Y'$ be an S-extension of Y by a quasi-coherent Module J . Let G be a commutative group scheme over S. Using the result mentioned at the end of (2.9), it can be shown that there is a canonical, functorial isomorphism of $D(\mathbb{Z}_{Y'})$:

(4.5.1) $$j_{*}(\mathcal{X}^{\vee}_{G_Y} \overset{L}{\otimes} J) \simeq (0 \to G_{Y'} \xrightarrow{d} j_{*}G_Y \to 0)$$

([1]) i.e. the group of invertible elements in $A^{\vee}$.

where $G_{Y'}$ is placed in degree 0 and d is the adjunction map ($D(\mathbb{Z}_{Y'})$ means the derived category of $\mathbb{Z}$-Modules on the large fpqc site of Y'). This formula was conjectured by Grothendieck after his reading [15], and proven by Deligne. It yields in particular a canonical, functorial description of the Lie complex $\ell_G^\vee$ as an object of $D(\mathbb{Z}_S)$, since we can take for j the inclusion of S into the scheme of dual numbers on $\underline{O}_S$. Observing that $R^1 j_*(G_Y) = 0$, we derive from (4.5.1) an exact sequence

$$\ldots \to H^i(Y',G_{Y'}) \longrightarrow H^i(Y,G_Y) \longrightarrow H^{i+1}(Y,\ell_{G_Y} \overset{L}{\otimes} J) \to \ldots$$

for $i \leq 1$, which shows again (cf. (2.6)[1]) that the obstruction to deforming a G_Y-torsor over Y' lies in $H^2(Y,\ell_{G_Y}^\vee \overset{L}{\otimes} J)$ (moreover, the above sequence can be interpreted as an exact sequence of relative cohomology (3.3), hence the other parts of (2.6)).

4.6. To conclude these generalities, let us mention a very striking formula for the Lie complex in the finite case, which is due to Grothendieck (see [11]). Let G be a finite, locally free, commutative group scheme over S, and G^* denote its Cartier dual. Let I be a quasi-coherent $\underline{O}_S$-Module. Then, there is a canonical, functorial isomorphism in D(S) :

(4.6.1) $$\ell_G^\vee \overset{L}{\otimes} I \simeq t_{1]}\, \underline{RHom}_{\mathbb{Z}}(G^*,I) \quad (\simeq t_{1]}\, \underline{RHom}_{\underline{O}_S}(G^* \overset{L}{\otimes}_{\mathbb{Z}} \underline{O}_S, I))$$

where $t_{n]}L$, for a complex L, means the truncated complex obtained by killing $H^i(L)$ for $i > n$, i.e. $t_{n]}L = (\ldots \to L^{n-2} \longrightarrow L^{n-1} \longrightarrow Z^n \longrightarrow 0)$. In particular, with the notations of (4.1), we deduce from (4.6.1) canonical, functorial isomorphisms :

(4.6.2) $$t_G \simeq \underline{Hom}(G^*,I) \quad , \quad \nu_G \simeq \underline{Ext}^1(G^*,I) \quad .$$

The above formulas are helpful in the study of the co-Lie and Lie complexes of truncated Barsotti-Tate groups. For details the reader is referred to [11], where he will also find interesting developments concerning the

---

(1) Note that (3.1.2) and (4.4) imply $\ell_{X/Y} \simeq \ell_{G_Y}$.

relationship between Lie complexes and Dieudonné modules.

5. Deformations of commutative, flat group schemes.

As in § 4, all our group schemes will be assumed to be flat and locally of finite presentation, unless otherwise stated. We fix a scheme S.

5.1. Let A be a ring scheme over S, not necessarily commutative, but associative and unitary. We don't assume the underlying scheme to be flat or locally of finite presentation. By an A-module scheme over S we mean a commutative group scheme G over S, endowed with an A-module structure, i.e. a bi-linear map $A \times_S G \longrightarrow G$, $(a,g) \mapsto ag$, such that $a(bg) = (ab)g$, $1g = g$, for any T-valued points a, b of A, g of G. We have especially in mind the case where A is the constant ring scheme $\mathbb{Z}_S$ (resp. $(\mathbb{Z}/n\mathbb{Z})_S$), in which case an A-module scheme is simply a commutative group scheme (resp. a commutative group scheme killed by n). But other cases may be of interest, e. g. $A = \mathbb{Z}_S[\Gamma]$ (Γ being a discrete group or monoïd), $A = \underline{O}_S$, $A = \underline{\underline{W}}_S$, the universal Witt scheme over S ([17] p. 179).

Let G be an A-module scheme over S. Thanks to the action of A, it is possible to define a finer object than ℓ_G, namely an object

(5.1.1) $$\ell_G^A \in \text{ob } D(A \otimes_{\mathbb{Z}} \underline{O}_S)$$

whose image under the forgetful functor $D(A \otimes_{\mathbb{Z}} \underline{O}_S) \longrightarrow D(S)$ is canonically isomorphic to ℓ_G. If G is smooth, we can take for ℓ_G^A the sheaf of invariant differential forms ω_G endowed with its natural A-linear structure. The definition of (5.1.1) in full generality is a little sophisticated (as were the definitions of $L^G_{X/Y}$ (2.1.1) or $\underline{\ell}_G$ (3.1.1)). One method consists in interpreting $\ell_G^{\vee}[1]$ as the stack of equivariant deformations of G over the dual numbers on S (4.5) and observing that the latter stack has a natural $A \otimes_{\mathbb{Z}} \underline{O}_S$-linear structure.

hence defines, by the dictionary of (SGA 4 XVIII), a complex of $A \otimes_{\mathbb{Z}} \underline{O}_S$-Modules of length 1, which is $R\underline{Hom}_{\underline{O}_S}(\chi_G^A, \underline{O}_S)[1]$. Another method, in the style of (2.8 c)), uses a large diagram describing the structure of A-module of G (see [13] and (5.8) below).

We shall sometimes write χ_G instead of χ_G^A. The notation $\chi_G^\vee$ will mean $R\underline{Hom}_{\underline{O}_S}(\chi_G^A, \underline{O}_S)$ ($\in$ ob $D(A \otimes_{\mathbb{Z}} \underline{O}_S)$).

Some of the results of § 4 admit natural refinements. For instance, an exact sequence (4.3.1) of A-module schemes gives rise to an exact triangle (4.3.2) in $D(A \otimes_{\mathbb{Z}} \underline{O}_S)$, and the isomorphism (4.5.1), for an A-module scheme G, comes from an isomorphism in D(A).

5.2. We suppose now S is over some fixed scheme T, and we are given a T-extension $i : S \to S'$ by a quasi-coherent $\underline{O}_S$-Module I. We suppose, moreover, that A is induced by a flat ring scheme A' over S', $A = A' \times_{S'} S$. Concerning the deformations of A-module schemes, our main result is the following :

<u>Theorem</u> 5.3. <u>In the situation of</u> (5.2), <u>let</u> G <u>be an</u> A-<u>module scheme over</u> S. <u>There is an obstruction</u>

$$\omega(G,i) \in \mathrm{Ext}^2_A(G, \chi_G^\vee \overset{L}{\otimes} I) \qquad (^1)$$

<u>whose vanishing is necessary and sufficient for the existence of an</u> A'-<u>module scheme</u> G' <u>over</u> S' <u>deforming</u> G, i.e. <u>equipped with an isomorphism of</u> A-<u>module schemes</u> $G' \times_{S'} S \xrightarrow{\sim} G$. <u>When</u> $\omega(G,i) = 0$, <u>the set of isomorphism classes of deformations</u> G' <u>is an affine space under</u> $\mathrm{Ext}^1_A(G, \chi_G^\vee \overset{L}{\otimes} I)$, <u>and the group of automorphisms of a given deformation is canonically isomorphic to</u> $\mathrm{Ext}^0_A(G, \chi_G^\vee \overset{L}{\otimes} I)$. <u>Moreover, if</u> A, A' <u>are induced by a fixed, flat ring scheme</u> B <u>over</u> T, <u>there is defined a canonical class, depending only on</u> B <u>and the composition</u> $G \to S \to T$,

$$c(B,G/S/T) \in \mathrm{Ext}^1_A(G, \chi_G^\vee \overset{L}{\otimes} tL_{S/T}) \qquad ,$$

---

(1) Ext are taken with respect to the fpqc topology on the category of schemes over S. For the notation $tL_{S/T}$ below, see (2.7).

such that $\omega(G,i)$ *is given by the cup-product*

$$\omega(G,i) = e(i)c(B,G/S/T)$$

where $e(i) \in \mathrm{Ext}^1(L_{S/T},I)$ *is the class of* i .

We also have two results concerning the deformations of morphisms of A-module schemes :

Theorem 5.4. *Let* F', G' *be* A'*-module schemes over* S', *and let* $f : F \longrightarrow G$ *be a morphism of* A*-module schemes, where* $F = F' \times_{S'} S$, $G = G' \times_{S'} S$. *There is a canonical obstruction, depending functorially on* F', G', f :

$$\omega(F',G',f,i) \in \mathrm{Ext}^1_A(F,\chi_G^\vee \overset{L}{\otimes} I)$$

whose vanishing is necessary and sufficient for the existence of a morphism of A'*-module schemes* $f' : F' \to G'$ *such that* $f' \times_{S'} S = f$. *When* $\omega(F',G',f,i) = 0$, *the set of solutions* f' *is an affine space under* $\mathrm{Ext}^0_A(F,\chi_G^\vee \overset{L}{\otimes} I)$.

Theorem 5.5. *Let* F' *be an* A'*-module scheme over* S', G *an* A*-module scheme over* S, *and* $f : F \longrightarrow G$ *a morphism of* A*-module schemes, where* $F = F' \times_{S'} S$. *Denote by* $C(f) \in \mathrm{ob}\, D(A)$ *the mapping-cylinder of* f . *There exists a canonical obstruction, depending functorially on* F', f :

$$\omega(F',f,i) \in \mathrm{Ext}^2_A(C(f),\chi_G^\vee \overset{L}{\otimes} I)$$

whose vanishing is necessary and sufficient for the ~~simultaneous~~ existence of an A'*-module scheme* G' *deforming* G *in the sense of* (5.3) *and a morphism of* A'*-module schemes* $f' : F' \longrightarrow G'$ *such that* $f' \times_{S'} S = f$. *When* $\omega(F',f,i) = 0$, *the set of isomorphism classes of solutions* (G',f') *is an affine space under* $\mathrm{Ext}^1_A(C(f),\chi_G^\vee \overset{L}{\otimes} I)$ *and the group of automorphisms of a given solution is canonically isomorphic to* $\mathrm{Ext}^0_A(C(f),\chi_G^\vee \overset{L}{\otimes} I)$.

5.6. As we said in the introduction, (5.3) and (5.4) were conjectured by Grothendieck (letter to the author, 12/2/69) and form a basic tool in the study of Barsotti-Tate groups on a general base ([11] and [16]). As for (5.5), it came from an attempt to prove by means of our theory an unpublished [1] result of Oort, that was kindly brought to our attention by Mazur, and says the following : in the situation of (5.5), suppose S' is affine, $A' = \mathbb{Z}_{S'}$, F' is finite over S', G is an abelian scheme over S, f is a closed immersion, then the obstruction $\omega(F',f,i)$ vanishes, in other words there exist an abelian scheme G' on S' lifting G and an embedding $f' : F' \to G'$ lifting f . Note that in this situation $C(f) = G/F$ is an abelian scheme. It should be the case, at least when 2 is invertible on S, that the Ext^2 of an abelian scheme with a locally free sheaf of finite rank is zero. If this is true, then Oort's result follows from (5.5).

Among (5.3), (5.4), (5.5) there are some compatibilities that we should like to discuss briefly.

a) In the situation of (5.4), suppose F = G and f is the identity. Then $\omega(F',G',f,i) \in \mathrm{Ext}^1_A(G, \ell^\vee_G \overset{L}{\otimes} I)$ is the difference, in the sense of (5.3), between the classes of the deformations F', G' of G .

b) In the situation of (5.5), suppose there exists an A'-module scheme G' deforming G . Then it follows from (5.3) that (F',f,i) lies in

$$\mathrm{Coker}(\mathrm{Ext}^1_A(G,M) \xrightarrow{f^*} \mathrm{Ext}^1_A(F,M)) \hookrightarrow \mathrm{Ext}^2_A(C(f),M)$$

where $M = \ell^\vee_G \overset{L}{\otimes} I$ and f^* is the map induced by f . It may be observed that the above cokernel corresponds to the obstruction to lifting f into a map $f' : F' \longrightarrow G'$ (5.4) modulo the indeterminacy (5.3) in the choice of G' .

5.7. Before we turn to the proof of the above results, we shall mention a reassuring compatibility between the obstruction $\omega(G,i)$ of (5.3) for A, A' = the constant ring scheme $\mathbb{Z}$ (obstruction to deforming G as a commutative group scheme) and the obstruction $\omega(G,i)$ of (3.4) (obstruction to deforming G as a (possibly non-commutative) group scheme. Recall that, for $M \in \mathrm{ob}\, D^+(\mathbb{Z}_S)$, there is a canonical, functorial map

$$\mathrm{RHom}_{\mathbb{Z}}(G,M) \longrightarrow R\Gamma(BG/S,p^*M)[1]$$

(where $p : BG \to S$ is the projection), which is defined by identifying $R\Gamma(BG/S,p^*M)$ with $\mathrm{RHom}_{\mathbb{Z}}(\mathbb{Z}(\mathrm{Ner}(G))/\mathbb{Z}, M)$ (where $\mathbb{Z}(-)$ denotes the free

[1] to our knowledge

abelian group functor) and using the canonical epimorphism $\mathbb{Z}(G) \to G$, which defines a map $\mathbb{Z}(\mathrm{Ner}(G))/\mathbb{Z} \longrightarrow G[1]$. In particular, thanks to (4.4) we have a map

$$\mathrm{RHom}_{\mathbb{Z}}(G, \ell_G^{\vee} \overset{L}{\otimes} I) \longrightarrow R\Gamma(BG/S, \ell_G^{\vee} \overset{L}{\otimes} I)[1] \quad ,$$

hence a map

$$(*) \qquad \mathrm{Ext}^{*}_{\mathbb{Z}}(G, \ell_G^{\vee} \overset{L}{\otimes} I) \longrightarrow H^{*+1}(BG/S, \ell_G^{\vee} \overset{L}{\otimes} I) \quad .$$

The compatibility says that the image of the obstruction $\omega(G,i)$ of (5.3) under (*) is the obstruction $\omega(G,i)$ of (3.4).

5.8. <u>Sketch of proof</u>. As in (2.8) and (3.6), the idea is to reduce the problem to a problem of deformation of a suitable ringed topos or map of ringed topoi. Thus (5.3) will be reduced to (1.7), while (5.4) and (5.5) will be reduced to variants of (1.7), namely ([12] III 2.2.4) and ([12] III 2.3.2) respectively. Since the technique is the same in the three cases, we shall restrict ourselves to (5.3). The proof is long and rather involved. We shall only outline the main steps. For details, the reader is again referred to [13] .

5.8.1. <u>Diagrams</u>. Let T be a category. By a diagram of T of type I we mean a functor $X : I \longrightarrow T$. Let X, Y be diagrams of T of types I, J respectively. By definition a map $f : X \to Y$ is a pair (u,v) where $v : I \to J$ is a functor and $u : X \to Yv$ is a functor morphism. Thus the diagrams of T form a category denoted by $\mathrm{Diagr}_1(T)$. We define $\mathrm{Diagr}_n(T)$, for $n \geqslant 0$, by the formulas :

$$\mathrm{Diagr}_o(T) = T \quad , \quad \mathrm{Diagr}_n(T) = \mathrm{Diagr}(\mathrm{Diagr}_{n-1}(T)) \quad .$$

The category $\mathrm{Diagr}_n(T)$ is called the <u>category of</u> n-<u>diagrams</u> of T . Observe that if T possesses finite products, the same is true for Diagr(T), hence for $\mathrm{Diagr}_n(T)$: if $X : I \to T$, $Y : J \to T$ are diagrams, then $X \times Y$ is the diagram of type $I \times J$ defined by $(i,j) \mapsto X_i \times Y_j$.

5.8.2. <u>Spectra</u>. Let T be a topos. As we have seen above, the nerve functor gives an embedding of the category of groups of T into the category of simplicial objects of T :

$$\mathrm{Ner} : \mathrm{Group}(T) \hookrightarrow \mathrm{Simpl}(T) \quad ,$$

whose essential image consists exactly of those simplicial objects which satisfy the exactness properties (i), (ii), (iii) of (3.6 a)) (Y replaced by the final object of T). Now, if G is a commutative group of T, the multiplication $G \times G \to G$ is a group homomorphism, hence Ner(G) is actually a simplicial object in the category of commutative groups of T, or equivalently a commutative group of the topos Simpl(T), therefore we can iterate the nerve construction and define Ner(Ner(G)), Ner(Ner(Ner(G))), etc. Denote by $\mathbb{Z}$-Mod(T) the category of commutative groups of T, and by n-Simpl(T), for $n \in \mathbb{N}$, the category of n-simplicial objects of T (i.e. the category of functors from $\Delta^{o} \times \dots \times \Delta^{o}$ (n times) to T where Δ is the category of finite, non void, totally ordered sets). For $G \in \mathrm{ob}\ \mathbb{Z}\text{-Mod}(T)$, we define inductively $G\langle n\rangle \in \mathrm{ob}$ n-Simpl(T) by

(5.8.2.1) $\quad G\langle 0\rangle = G \quad , \quad G\langle n\rangle = \mathrm{Ner}(G\langle n-1\rangle) \quad \text{for } n \geqslant 1 \ .$

The functor

(5.8.2.2) $\quad \mathbb{Z}\text{-}\mathrm{Mod}(T) \longrightarrow n\text{-}\mathrm{Simpl}(T) \quad , \quad G \longmapsto G\langle n\rangle \quad ,$

is faithful for n = 0, fully faithful for $n \geqslant 1$, and, for $n \geqslant 2$, its essential image consists exactly of those n-simplicial objects Y possessing the following property : the simplicial objects deduced from Y by fixing all variables but one satisfy the conditions (i), (ii), (iii) recalled above (this last assertion is an easy consequence of the well known fact that a group in the category of groups is a commutative group).

Let G be a commutative group of T . Observe that $G\langle n\rangle$ corresponds, by the normalization functor [8] , to the n-complex concentrated in degree (-1,...,-1) with value G. In particular, we can identify $G\langle n\rangle$ with each of the "faces" $G\langle n+1\rangle(*,\dots,[1],*,\dots,*)$, and we get an augmented, strictly cosimplicial object in Diagr(T) :

(5.8.2.3) $\quad \underline{\underline{G}} = (G \to G\langle 1\rangle \rightrightarrows G\langle 2\rangle \ \dots \ G\langle n\rangle \underset{n+1 \text{ arrows}}{\overset{\longrightarrow}{\underset{\longrightarrow}{\cdots}}} G\langle n+1\rangle \ \dots)$

which we shall call the spectral diagram (or, simply, spectrum) of G, by analogy with the Eilenberg-Mac-Lane spectra.

Let us return to (5.8.1) for a moment. To any n-diagram X of T is associated an (n-1)-diagram of (Cat) (the category of categories), called the type of X , and denoted by Typ(X). It depends functorially on X and is defined inductively by : Typ(X) = I if $X : I \to T$ is a diagram of T, and, if $X : I \to \mathrm{Diagr}_{n-1}(T)$ is an n-diagram with $n \geqslant 2$, Typ(X) is the diagram $i \mapsto \mathrm{Typ}(X_i)$. If t is a type of n-diagram (i.e. an (n-1)-diagram of (Cat)), we define the category of diagrams of type t, $\mathrm{Diagr}_t(T)$, as the category whose objects are the diagrams of type t and maps are the maps of diagrams inducing the identity on t . Denote by

$$(5.8.2.4)\qquad \mathbb{D} = (pt \to \Delta^o \rightrightarrows (\Delta^o)^2 \;\ldots\; (\Delta^o)^n \;\overset{\longrightarrow}{\underset{\longrightarrow}{\cdots}}\; (\Delta^o)^{n+1} \;\ldots)$$
$$(x_1,\ldots,x_n) \mapsto (x_1,\ldots,[1],\ldots,x_n)$$

the type of any spectral diagram $\underline{\underline{G}}$. We have an embedding

$$(5.8.2.5)\qquad \mathbb{Z}\text{-Mod}(T) \hookrightarrow \mathrm{Diagr}_{\mathbb{D}}(T) \quad , \quad G \mapsto \underline{\underline{G}} \quad ,$$

whose essential image consists exactly of the diagrams Y of type $\mathbb{D}$ such that, for each $n \in \mathbb{N}$, Y_n is in the essential image of (5.8.2.2) and the maps of n-simplicial objects induced by the n+1 maps $Y_n \longrightarrow Y_{n+1}$ are isomorphisms.

As the normalization functor commutes with external tensor products, we get, for X, $Y \in \mathrm{ob}\; \mathbb{Z}\text{-Mod}(T)$, p, $q \in \mathbb{N}$, a canonical map

$$(5.8.2.6)\qquad X\langle p\rangle \times Y\langle q\rangle \longrightarrow (X \otimes Y)\langle p+q\rangle \quad ,$$

which is associative in the obvious sense. Let A be a ring object in T (associative and unitary), and let G be a left A-Module. Then, thanks to the maps (5.8.2.6), $\underline{\underline{A}}$ becomes a monoïd object in $\mathrm{Diagr}_2(T)$ (associative and unitary), and $\underline{\underline{G}}$ becomes a (left) $\underline{\underline{A}}$-object. In particular, we can consider the nerves

$$\mathrm{Ner}(\underline{\underline{A}}) = (\ldots\ \underline{\underline{A}} \times \underline{\underline{A}} \ \substack{\rightarrow\\\rightarrow\\\rightarrow}\ \underline{\underline{A}} \rightrightarrows e) \qquad (e = \text{the final object of } T)\ ,$$

$$\mathrm{Ner}(\underline{\underline{A}},\underline{\underline{G}}) = (\ldots\ \underline{\underline{A}} \times \underline{\underline{A}} \times \underline{\underline{G}} \ \substack{\rightarrow\\\rightarrow\\\rightarrow}\ \underline{\underline{A}} \times \underline{\underline{G}} \rightrightarrows \underline{\underline{G}}) \qquad ,$$

which are certain objects of $\mathrm{Diagr}_3(T)$. There is a natural projection

(5.8.2.7) $\quad \mathrm{Ner}(\underline{\underline{A}},\underline{\underline{G}}) \longrightarrow \mathrm{Ner}(\underline{\underline{A}}) \quad ,$

and, as above, it is again possible to characterize A-Mod(T), the category of A-Modules of T, as a certain category of diagrams over $\mathrm{Ner}(\underline{\underline{A}})$ satisfying certain exactness conditions.

5.8.3. Reduction to (1.7). Applying the above to the situation of (5.3), we get a diagram

$$\begin{array}{ccc} \mathrm{Ner}(\underline{\underline{A}},\underline{\underline{G}}) & & \\ {\scriptstyle f}\downarrow & & \\ \mathrm{Ner}(\underline{\underline{A}}) & \hookrightarrow & \mathrm{Ner}(\underline{\underline{A}}') \end{array} \quad ,$$

which in turn can be interpreted as a diagram of ringed topoi (like in (2.8)), in which the horizontal map is a T-extension of $\mathrm{Ner}(\underline{\underline{A}})$ by the inverse image of I, still denoted by I . From the generalities of (5.8.2) and the flatness criterion it follows that an A'-module scheme G' deforming G is essentially the same as a deformation of f over $\mathrm{Ner}(\underline{\underline{A}}')$ as a map of ringed topoi. Therefore we can apply (1.7), and it remains to calculate the groups $\mathrm{Ext}^i(L_{\mathrm{Ner}(\underline{\underline{A}},\underline{\underline{G}})/\mathrm{Ner}(\underline{\underline{A}})}, f^*I)$ (or, more generally, the analogous groups with I replaced by a complex of $\underline{O}_S$-Modules with bounded, quasi-coherent cohomology, e. g. a truncation of $L_{S/T}$). This is, however, far from being easy. We shall briefly indicate the main points.

5.8.4. A duality formula. Fix a topos T . To a diagram $X : I \longrightarrow T$ there are associated two topoi : Top(X), $\mathrm{Top}^{\circ}(X)$. The first one is the total topos of (SGA 4 VI) defined by the fibered topos $i \mapsto T_{/X_i}$; its objects are families E consisting of a sheaf E_i on X_i for each object i of I together with a map $X_f^* E_j \longrightarrow E_i$ for each map $f : i \to j$ of I, these maps satisfying certain transitivity relations. The other one, $\mathrm{Top}^{\circ}(X)$, is defined in the same way, except that we reverse the sense of the transition arrows, namely give ourselves a map $E_i \to X_f^* E_j$ for each $f : i \to j$; in other words, $\mathrm{Top}^{\circ}(X)$ is the topos of diagrams of type I over X. The construction of Top(X), $\mathrm{Top}^{\circ}(X)$ easily extends to n-diagrams.

Fix a Ring $\underline{O}$ of T . If X is an n-diagram of T, we equip Top(X), $\mathrm{Top}^{\circ}(X)$ with the Rings induced by $\underline{O}$. Now, if L is a Module on Top(X) and M a Module on T, we can define a Module $\underline{\underline{\mathrm{Hom}}}^{!}(L,M)$ on $\mathrm{Top}^{\circ}(X)$ with the properties that it depends functorially on L, M and induces the ordinary $\underline{\underline{\mathrm{Hom}}}$ on each piece of the diagram. For example, if X is a 1-diagram $I \to T$, we define $\underline{\underline{\mathrm{Hom}}}^{!}(L,M)$ as the family $i \mapsto \underline{\underline{\mathrm{Hom}}}(L_i, M_{X_i})$ with the obvious transition maps, and this generalizes trivially to n-diagrams. Moreover, we can derive the construction of $\underline{\underline{\mathrm{Hom}}}^{!}$, namely define

$$R\underline{\underline{\mathrm{Hom}}}^{!}(L,M) \in \mathrm{ob}\ D(\mathrm{Top}^{\circ}(X))$$

as a bi-functor of $L \in \mathrm{ob}\ D(\mathrm{Top}(X))$, $M \in \mathrm{ob}\ D^{+}(T)$. It is not hard to prove the following "duality formula" :

<u>Proposition</u> 5.8.4.1. <u>With the above notations</u>, <u>there exists a canonical</u>, <u>functorial isomorphism</u>

$$\mathrm{RHom}(\mathrm{Top}(X);L,M_X) \xrightarrow{\sim} R\Gamma(\mathrm{Top}^{\circ}(X);R\underline{\underline{\mathrm{Hom}}}^{!}(L,M))$$

<u>for</u> $L \in \mathrm{ob}\ D^{-}(\mathrm{Top}(X))$, $M \in \mathrm{ob}\ D^{+}(T)$.

5.8.5. <u>Passing to the tangent complex</u>. Our goal is to calculate

$$(*) \qquad \mathrm{RHom}(L_{\mathrm{Ner}(\underline{\underline{A}},\underline{\underline{G}})/\mathrm{Ner}(\underline{\underline{A}})}, f^{*}M)$$

where M is a complex of $\underline{O}_S$-Modules with bounded, quasi-coherent cohomology. By definition (5.8.3), (*) is an RHom of complexes of $\underline{O}$-Modules on the total topos of $\mathrm{Ner}(\underline{\underline{A}},\underline{\underline{G}})$ obtained by associating to each scheme of the diagram its small Zariski topos. But, as both arguments inside the RHom have quasi-coherent cohomology, they can, in a natural way, be extended into objects of $D(\mathrm{Top}(\mathrm{Ner}(\underline{\underline{A}},\underline{\underline{G}})))$ where now $\mathrm{Ner}(\underline{\underline{A}},\underline{\underline{G}})$ is viewed as a 3-diagram in the (large) fpqc topos of S, and by descent the RHom does not change under this extension. Now we can apply (5.8.4.1) and we get an isomorphism

$$(**)\quad \mathrm{RHom}(L_{\mathrm{Ner}(\underline{\underline{A}},\underline{\underline{G}})/\mathrm{Ner}(\underline{\underline{A}})}, f^{*}M) \simeq R\Gamma(\mathrm{Top}^{o}(\mathrm{Ner}(\underline{\underline{A}},\underline{\underline{G}})), \underline{\mathrm{RHom}}^{!}(L_{\mathrm{Ner}(\underline{\underline{A}},\underline{\underline{G}})/\mathrm{Ner}(\underline{\underline{A}})}, M))$$

which brings (*) nearer to calculation, because $\underline{\mathrm{RHom}}^{!}(L_{\mathrm{Ner}(\underline{\underline{A}},\underline{\underline{G}})/\mathrm{Ner}(\underline{\underline{A}})}, M)$ has a very simple interpretation. In effect, denote by $\mathrm{Lie}(G) = \ell_G^{\vee}$ (5.1) the Lie complex of G, and by $p : \mathrm{Ner}(\underline{\underline{A}},\underline{\underline{G}}) \longrightarrow \mathrm{Ner}(\underline{\underline{A}},\underline{\underline{O}})$ the canonical map defined by projecting G to zero. The construction $G \mapsto \mathrm{Ner}(\underline{\underline{A}},\underline{\underline{G}})$ extends trivially to complexes of A-Modules, and it is easily deduced from the Mazur-Roberts exact triangle (4.3.2) that there is a canonical isomorphism of $D^{b}(\mathrm{Top}^{o}(\mathrm{Ner}(\underline{\underline{A}},\underline{\underline{G}})))$:

$$(***)\quad \underline{\mathrm{RHom}}^{!}(L_{\mathrm{Ner}(\underline{\underline{A}},\underline{\underline{G}})/\mathrm{Ner}(\underline{\underline{A}})}, M) \simeq p^{*}\mathrm{Ner}(\underline{\underline{A}},\underline{\underline{\underline{\mathrm{Lie}}}}(G)) \overset{L}{\otimes} f^{*}M$$

(the tensor product on the right being of course taken over the ring induced by $\underline{O}_S$). In view of (5.8.3), (5.3) will follow from the combination of (**), (***), and the following canonical isomorphism

$$R\Gamma(\mathrm{Top}^{o}(\mathrm{Ner}(\underline{\underline{A}},\underline{\underline{G}})), p^{*}\mathrm{Ner}(\underline{\underline{A}},\underline{\underline{\underline{\mathrm{Lie}}}}(G)) \overset{L}{\otimes} f^{*}M) \simeq \mathrm{RHom}_A(G, \mathrm{Lie}(G) \overset{L}{\otimes} M) \quad ,$$

which is itself a consequence of the more general

Theorem 5.8.6. <u>Fix a topos</u> T, <u>an associative and unitary ring</u> A <u>of</u> T, <u>a commutative</u> (<u>and unitary</u>) <u>ring</u> R <u>of</u> T, $G \in \mathrm{ob}\ A\text{-}\mathrm{Mod}(T)$, $E \in \mathrm{ob}\ D(A \otimes_{\mathbb{Z}} R)$, $M \in \mathrm{ob}\ D^{b}(R)$. <u>Assume</u> E <u>is of finite flat amplitude as a complex of</u> R-<u>Modules</u>. <u>Denote by</u> $p : \mathrm{Ner}(\underline{\underline{A}},\underline{\underline{G}}) \rightarrow \mathrm{Ner}(\underline{\underline{A}},\underline{\underline{O}})$, $f : \mathrm{Ner}(\underline{\underline{A}},\underline{\underline{G}}) \rightarrow T$ <u>the canonical projections</u>. <u>Then there exists a canonical</u>, <u>functorial isomorphism</u>

$$R\Gamma(\mathrm{Top}^{o}(\mathrm{Ner}(\underline{\underline{A}},\underline{\underline{G}})), p^{*}\mathrm{Ner}(\underline{\underline{A}},\underline{\underline{E}}) \overset{L}{\otimes}_R f^{*}M) \simeq \mathrm{RHom}_A(G, E \overset{L}{\otimes}_R M) \quad .$$

<u>Proof</u>. We shall just sketch the idea. For simplicity, we shall assume M = R . Denote by $\mathbb{Z}^{st}(-)$ the functor deduced by stabilization from the functor $X \mapsto \mathbb{Z}(X)$ on the category of abelian groups of T, where $\mathbb{Z}(X)$ means the free abelian group on the underlying sheaf of sets. For $X \in \mathrm{ob}\ \mathbb{Z}\text{-}\mathrm{Mod}(T)$, $\mathbb{Z}^{st}(X)$ is a simplicial abelian group of T, defined by

$$\mathbb{Z}^{st}(X) = \varinjlim \mathbb{Z}(X[n])[-n]$$

where the shifts of degrees are performed simplicially thanks to the Dold-Puppe equivalence, and the direct limit is taken with respect to the "suspension maps". Using the pairings (5.8.2.6), it is possible to turn $\mathbb{Z}^{st}(A)$ into an associative and unitary simplicial ring of T, and $\mathbb{Z}^{st}(G)$ into a $\mathbb{Z}^{st}(A)$-Module. Moreover, we have a canonical map of rings $\mathbb{Z}^{st}(A) \longrightarrow A$, and a $\mathbb{Z}^{st}(A)$-linear map $\mathbb{Z}^{st}(G) \to G$. Now, the proof of (5.8.6) breaks down into two parts :

a) First, using standard resolutions, we show there is a canonical isomorphism

$$R\Gamma(\mathrm{Top}^{\circ}(\mathrm{Ner}(\underline{\underline{A}},\underline{\underline{G}})),p^{*}\mathrm{Ner}(\underline{\underline{A}},\underline{\underline{E}})) \simeq \mathrm{RHom}_{A}(\mathbb{Z}^{st}(G) \overset{L}{\otimes}_{\mathbb{Z}^{st}(A)} A\ ,\ E) \quad .$$

(This is the hard part of the proof).

b) Second, we show that the canonical map

$$(5.8.6.1) \qquad \mathbb{Z}^{st}(G) \overset{L}{\otimes}_{\mathbb{Z}^{st}(A)} A \longrightarrow G$$

is an isomorphism. This last result is essentially due to Mac-Lane [14], as was explained to us by L. Breen. It is easy to prove by dévissage and reduction to the case G = A .

This concludes the (sketchy) proof of (5.8.6), and therefore demonstrates (5.3).

Remark 5.9. It is easy to deduce from (5.8.6.1) a canonical, functorial resolution of G of the form desired by Grothendieck in (SGA 7 VII 3.5.4).

Remark 5.10. Deligne has indicated a method based on the theory of Picard stacks that should yield another proof of the results of this section.

BIBLIOGRAPHY

[1] André, M., Méthode simpliciale en Algèbre Homologique et Algèbre Commutative, Lecture Notes in Mathematics 32, 1967.

[2] Artin, M., Grothendieck, A. et Verdier, J. L., Théorie des topos et cohomologie étale des schémas, Séminaire de Géométrie Algébrique du Bois-Marie, 1963-64, to appear in the Lecture Notes (quoted SGA 4).

[3] Atiyah, M. F., Analytic connexions on fibre bundles, Mexico Symposium, 1958.

[4] Berthelot, P., Grothendieck, A., et Illusie, L., Théorie des intersections et Théorème de Riemann-Roch, Séminaire de Géométrie Algébrique du Bois-Marie, 1966-67 (SGA 6), Lecture Notes in Mathematics 225, 1971.

[5] Breen, L., work in preparation (a preliminary version exists in the form of mimeographed notes (M. I. T., 1971) : On some extensions of abelian sheaves in dimensions two and three). See also : Extensions of Abelian Sheaves and Eilenberg-MacLane Algebras, Inventiones Math. 9, 15-44 (1969).

[6] Deligne, P. et Grothendieck, A., Le groupe de monodromie en Géométrie Algébrique, Séminaire de Géométrie Algébrique du Bois-Marie, 1968-69, quoted (SGA 7).

[7] Demazure, M. et Grothendieck, A., Schémas en groupes, Séminaire de Géométrie Algébrique du Bois-Marie, 1963-64 (SGA 3), Lecture Notes in Mathematics 151, 152, 153, 1970.

[8] Dold, A., and Puppe, D., Homologie nicht-additiver Funktoren, Anwendungen, Ann. Inst. Fourier, 11, 201-312 (1961).

[9] Giraud, J., Cohomologie non abélienne de degré 2, (Thèse), Die Grundlehren der mathematischen Wissenschaften 179, Springer-Verlag (1971).

[10] Grothendieck, A. et Dieudonné, J., Eléments de Géométrie Algébrique (EGA), pub. math. IHES, Paris.

[11] Grothendieck, A., Groupes de Barsotti-Tate, Cours au Collège de France, 1970-71 et 71-72, to appear (?).

[12] Illusie, L., Complexe cotangent et déformations I, Lecture Notes in Mathematics 239, 1971.

[13] Illusie, L., Complexe cotangent et déformations II, in preparation (to appear in the Lecture Notes).

[14] MacLane, S., Homologie des anneaux et des modules, Colloque de Topologie algébrique, Louvain, 55-80 (1956).

[15] Mazur, B., and Roberts, L., Local Euler Characteristics, Inv. Math. 9, 201-234 (1970).

[16] Messing, W., The Crystals associated to Barsotti-Tate Groups, with applications to Abelian Schemes, Thesis, Princeton (1971), to appear.

[17] Mumford, D., Lectures on Curves on an Algebraic Surface, Ann. of Math. Studies 59, Princeton (1966).

[18] Quillen, D. G., Notes on the homology of commutative rings, Mimeographed Notes, M. I. T. (1968), and On the (co-) homology of commutative rings, Proceedings of Symposia in Pure Mathematics 17, 65-87 (1970).

Lecture Notes in Mathematics

Comprehensive leaflet on request

Vol. 74: A. Fröhlich, Formal Groups. IV, 140 pages 1968 DM 16,-

Vol. 75: G. Lumer, Algèbres de fonctions et espaces de Hardy VI, 80 pages. 1968. DM 16,-

Vol. 76: R. G. Swan, Algebraic K-Theory. IV, 262 pages. 1968. DM 18,-

Vol. 77: P.-A. Meyer, Processus de Markov: la frontière de Martin. IV, 123 pages. 1968. DM 16,-

Vol. 78: H. Herrlich, Topologische Reflexionen und Coreflexionen. XVI, 166 Seiten. 1968. DM 16,-

Vol. 79: A. Grothendieck, Catégories Cofibrées Additives et Complexe Cotangent Relatif. IV, 167 pages. 1968. DM 16,-

Vol. 80: Seminar on Triples and Categorical Homology Theory Edited by B. Eckmann. IV, 398 pages. 1969. DM 20,-

Vol. 81: J.-P. Eckmann et M. Guenin, Méthodes Algébriques en Mécanique Statistique. VI, 131 pages. 1969. DM 16,-

Vol. 82: J. Wloka, Grundräume und verallgemeinerte Funktionen. VIII, 131 Seiten. 1969. DM 16,-

Vol. 83: O. Zariski, An Introduction to the Theory of Algebraic Surfaces. Second Printing. IV, 100 pages. 1972. DM 16,-

Vol. 84: H. Lüneburg, Transitive Erweiterungen endlicher Permutationsgruppen. IV, 119 Seiten. 1969. DM 16,-

Vol. 85: P. Cartier et D. Foata, Problèmes combinatoires de commutation et réarrangements. IV, 88 pages. 1969. DM 16,-

Vol. 86: Category Theory, Homology Theory and their Applications I. Edited by P. Hilton. VI, 216 pages. 1969. DM 16,-

Vol. 87: M. Tierney, Categorical Constructions in Stable Homotopy Theory. IV, 65 pages. 1969. DM 16,-

Vol. 88: Séminaire de Probabilités III IV, 229 pages. 1969. DM 18,-

Vol. 89: Probability and Information Theory. Edited by M. Behara, K. Krickeberg and J. Wolfowitz. IV, 256 pages. 1969. DM 18,-

Vol. 90: N. P. Bhatia and O. Hajek, Local Semi-Dynamical Systems. II, 157 pages. 1969. DM 16,-

Vol. 91: N. N. Janenko, Die Zwischenschrittmethode zur Lösung mehrdimensionaler Probleme der mathematischen Physik. VIII, 194 Seiten. 1969. DM 16,80

Vol. 92: Category Theory, Homology Theory and their Applications II Edited by P. Hilton. V, 308 pages. 1969. DM 20,-

Vol. 93: K. R. Parthasarathy, Multipliers on Locally Compact Groups. III, 54 pages. 1969. DM 16,-

Vol. 94: M Machover and J. Hirschfeld, Lectures on Non-Standard Analysis. VI, 79 pages. 1969 DM 16,-

Vol 95: A. S. Troelstra, Principles of Intuitionism II, 111 pages 1969 DM 16,-

Vol. 96: H.-B. Brinkmann und D. Puppe, Abelsche und exakte Kategorien, Korrespondenzen V, 141 Seiten. 1969. DM 16,-

Vol. 97: S. O Chase and M. E. Sweedler, Hopf Algebras and Galois theory. II, 133 pages. 1969. DM 16,-

Vol. 98: M. Heins, Hardy Classes on Riemann Surfaces. III, 106 pages 1969. DM 16,-

Vol. 99: Category Theory, Homology Theory and their Applications III Edited by P. Hilton. IV, 489 pages 1969. DM 24,-

Vol. 100: M. Artin and B. Mazur, Etale Homotopy. II, 196 Seiten 1969. DM 16,-

Vol. 101: G. P. Szegö et G. Treccani, Semigruppi di Trasformazioni Multivoche. VI, 177 pages. 1969. DM 16,-

Vol. 102: F. Stummel, Rand- und Eigenwertaufgaben in Sobolewschen Räumen. VIII, 386 Seiten. 1969. DM 20,-

Vol. 103: Lectures in Modern Analysis and Applications I Edited by C. T. Taam. VII, 162 pages. 1969. DM 16,-

Vol. 104: G. H. Pimbley, Jr., Eigenfunction Branches of Nonlinear Operators and their Bifurcations. II, 128 pages 1969. DM 16,-

Vol. 105: R. Larsen, The Multiplier Problem VII, 284 pages 1969. DM 18,-

Vol. 106: Reports of the Midwest Category Seminar III Edited by S. Mac Lane. III, 247 pages. 1969. DM 16,

Vol. 107: A. Peyerimhoff, Lectures on Summability III, 111 pages 1969. DM 16,-

Vol. 108: Algebraic K-Theory and its Geometric Applications. Edited by R. M.F. Moss and C. B. Thomas. IV, 86 pages. 1969. DM 16,-

Vol 109: Conference on the Numerical Solution of Differential Equations. Edited by J. Ll. Morris. VI, 275 pages 1969. DM 18,-

Vol. 110: The Many Facets of Graph Theory. Edited by G Chartrand and S. F. Kapoor. VIII, 290 pages. 1969. DM 18,-

Vol. 111: K H Mayer, Relationen zwischen charakteristischen Zahlen III, 99 Seiten 1969 DM 16,-

Vol. 112: Colloquium on Methods of Optimization Edited by N. N Moiseev. IV, 293 pages. 1970. DM 18,-

Vol 113: R Wille, Kongruenzklassengeometrien III, 99 Seiten 1970 DM 16,-

Vol. 114: H Jacquet and R P. Langlands, Automorphic Forms on GL (2) VII, 548 pages 1970 DM 24,-

Vol. 115: K H. Roggenkamp and V Huber-Dyson, Lattices over Orders I. XIX, 290 pages. 1970 DM 18,-

Vol 116: Séminaire Pierre Lelong (Analyse) Année 1969. IV, 195 pages 1970 DM 16,-

Vol. 117: Y Meyer, Nombres de Pisot, Nombres de Salem et Analyse Harmonique 63 pages. 1970. DM 16,-

Vol. 118: Proceedings of the 15th Scandinavian Congress, Oslo 1968 Edited by K. E. Aubert and W. Ljunggren. IV, 162 pages. 1970. DM 16,-

Vol. 119: M. Raynaud, Faisceaux amples sur les schémas en groupes et les espaces homogènes. III, 219 pages. 1970 DM 16,-

Vol. 120: D. Siefkes, Büchi's Monadic Second Order Successor Arithmetic XII, 130 Seiten 1970 DM 16,-

Vol. 121: H S Bear, Lectures on Gleason Parts. III, 47 pages. 1970 DM 16,-

Vol. 122: H. Zieschang, E. Vogt und H.-D. Coldewey, Flächen und ebene diskontinuierliche Gruppen. VIII, 203 Seiten. 1970. DM 16,-

Vol 123: A V Jategaonkar, Left Principal Ideal Rings. VI, 145 pages. 1970. DM 16,-

Vol 124: Séminare de Probabilités IV Edited by P. A Meyer IV, 282 pages 1970. DM 20,-

Vol 125: Symposium on Automatic Demonstration V, 310 pages 1970. DM 20,-

Vol. 126: P Schapira, Théorie des Hyperfonctions. XI, 157 pages 1970. DM 16,-

Vol 127: I Stewart, Lie Algebras IV, 97 pages 1970 DM 16,-

Vol 128: M Takesaki, Tomita's Theory of Modular Hilbert Algebras and its Applications. II, 123 pages. 1970 DM 16,-

Vol. 129: K. H. Hofmann, The Duality of Compact Semigroups and C*-Bigebras XII, 142 pages. 1970. DM 16,-

Vol 130: F Lorenz, Quadratische Formen über Körpern. II, 77 Seiten. 1970. DM 16,-

Vol. 131: A Borel et al., Seminar on Algebraic Groups and Related Finite Groups. VII, 321 pages. 1970 DM 22,-

Vol 132: Symposium on Optimization. III, 348 pages 1970 DM 22,-

Vol. 133: F Topsøe, Topology and Measure. XIV, 79 pages 1970 DM 16,-

Vol. 134: L. Smith, Lectures on the Eilenberg-Moore Spectral Sequence. VII, 142 pages 1970 DM 16,-

Vol. 135: W Stoll, Value Distribution of Holomorphic Maps into Compact Complex Manifolds II, 267 pages 1970 DM 18,-

Vol. 136: M. Karoubi et al., Séminaire Heidelberg-Saarbrücken-Strasbourg sur la K-Théorie. IV, 264 pages. 1970. DM 18,-

Vol. 137: Reports of the Midwest Category Seminar IV. Edited by S MacLane. III, 139 pages. 1970. DM 16,-

Vol 138: D Foata et M. Schützenberger, Théorie Géométrique des Polynômes Eulériens. V, 94 pages. 1970. DM 16,-

Vol. 139: A. Badrikian, Séminaire sur les Fonctions Aléatoires Linéaires et les Mesures Cylindriques VII, 221 pages. 1970 DM 18,-

Vol. 140: Lectures in Modern Analysis and Applications II. Edited by C. T. Taam. VI, 119 pages 1970. DM 16,-

Vol. 141: G. Jameson, Ordered Linear Spaces. XV, 194 pages. 1970 DM 16,-

Vol. 142: K. W Roggenkamp, Lattices over Orders II. V, 388 pages. 1970. DM 22,-

Vol. 143: K. W. Gruenberg, Cohomological Topics in Group Theory XIV, 275 pages 1970. DM 20,-

Vol. 144: Seminar on Differential Equations and Dynamical Systems, II. Edited by J A. Yorke. VIII, 268 pages. 1970 DM 20,-

Vol. 145: E. J. Dubuc, Kan Extensions in Enriched Category Theory. XVI, 173 pages. 1970 DM 16,-

Please turn over

Vol. 146: A. B. Altman and S. Kleiman, Introduction to Grothendieck Duality Theory. II, 192 pages. 1970. DM 18,–

Vol. 147: D. E. Dobbs, Cech Cohomological Dimensions for Commutative Rings. VI, 176 pages. 1970. DM 16,–

Vol. 148: R. Azencott, Espaces de Poisson des Groupes Localement Compacts. IX, 141 pages. 1970. DM 16,–

Vol. 149: R. G. Swan and E. G. Evans, K-Theory of Finite Groups and Orders. IV, 237 pages. 1970. DM 20,–

Vol. 150: Heyer, Dualität lokalkompakter Gruppen. XIII, 372 Seiten. 1970. DM 20,–

Vol. 151: M. Demazure et A. Grothendieck, Schémas en Groupes I. (SGA 3). XV, 562 pages. 1970. DM 24,–

Vol. 152: M. Demazure et A. Grothendieck, Schémas en Groupes II (SGA 3). IX, 654 pages. 1970. DM 24,–

Vol. 153: M. Demazure et A. Grothendieck, Schémas en Groupes III (SGA 3). VIII, 529 pages. 1970. DM 24,–

Vol. 154: A. Lascoux et M. Berger, Variétés Kähleriennes Compactes. VII, 83 pages. 1970. DM 16,–

Vol. 155: Several Complex Variables I, Maryland 1970. Edited by J. Horváth. IV, 214 pages. 1970. DM 18,–

Vol. 156: R. Hartshorne, Ample Subvarieties of Algebraic Varieties. XIV, 256 pages. 1970. DM 20,–

Vol. 157: T. tom Dieck, K. H. Kamps und D. Puppe, Homotopietheorie. VI, 265 Seiten. 1970. DM 20,–

Vol. 158: T. G. Ostrom, Finite Translation Planes. IV. 112 pages. 1970. DM 16,–

Vol. 159: R. Ansorge und R. Hass. Konvergenz von Differenzenverfahren für lineare und nichtlineare Anfangswertaufgaben. VIII, 145 Seiten. 1970. DM 16,–

Vol. 160: L. Sucheston, Constributions to Ergodic Theory and Probability. VII, 277 pages. 1970. DM 20,–

Vol. 161: J. Stasheff, H-Spaces from a Homotopy Point of View. VI, 95 pages. 1970. DM 16,–

Vol. 162: Harish-Chandra and van Dijk, Harmonic Analysis on Reductive p-adic Groups. IV, 125 pages. 1970. DM 16,–

Vol. 163: P. Deligne, Equations Différentielles à Points Singuliers Reguliers. III, 133 pages. 1970. DM 16,–

Vol. 164: J. P. Ferrier, Seminaire sur les Algebres Complètes. II, 69 pages. 1970. DM 16,–

Vol. 165: J. M. Cohen, Stable Homotopy. V, 194 pages. 1970. DM 16.

Vol. 166: A. J. Silberger, PGL_2 over the p-adics: its Representations, Spherical Functions, and Fourier Analysis. VII, 202 pages. 1970. DM 18,–

Vol. 167: Lavrentiev, Romanov and Vasiliev, Multidimensional Inverse Problems for Differential Equations. V, 59 pages. 1970. DM 16,–

Vol. 168: F. P. Peterson, The Steenrod Algebra and its Applications: A conference to Celebrate N. E. Steenrod's Sixtieth Birthday. VII, 317 pages. 1970. DM 22,–

Vol. 169: M. Raynaud, Anneaux Locaux Henséliens. V, 129 pages. 1970. DM 16,–

Vol. 170: Lectures in Modern Analysis and Applications III. Edited by C. T. Taam. VI, 213 pages. 1970. DM 18,–

Vol. 171: Set-Valued Mappings, Selections and Topological Properties of 2^X. Edited by W. M. Fleischman. X, 110 pages. 1970. DM 16,–

Vol. 172: Y.-T. Siu and G. Trautmann, Gap-Sheaves and Extension of Coherent Analytic Subsheaves. V, 172 pages. 1971. DM 16,–

Vol. 173: J. N. Mordeson and B. Vinograde, Structure of Arbitrary Purely Inséparable Extension Fields. IV, 138 pages. 1970. DM 16,–

Vol. 174: B. Iversen, Linear Determinants with Applications to the Picard Scheme of a Family of Algebraic Curves. VI, 69 pages. 1970. DM 16,–

Vol. 175: M. Brelot, On Topologies and Boundaries in Potential Theory. VI, 176 pages. 1971. DM 18,–

Vol. 176: H. Popp, Fundamentalgruppen algebraischer Mannigfaltigkeiten. IV, 154 Seiten. 1970. DM 16,

Vol. 177: J. Lambek, Torsion Theories, Additive Semantics and Rings of Quotients. VI, 94 pages. 1971. DM 16,–

Vol. 178: Th. Bröcker und T. tom Dieck, Kobordismentheorie. XVI, 191 Seiten. 1970. DM 18,–

Vol. 179: Seminaire Bourbaki – vol. 1968/69. Exposés 347-363. IV. 295 pages. 1971. DM 22,–

Vol. 180: Séminaire Bourbaki – vol. 1969/70. Exposés 364-381. IV, 310 pages. 1971. DM 22,–

Vol. 181: F. DeMeyer and E. Ingraham, Separable Algebras over Commutative Rings. V, 157 pages. 1971. DM 16,–

Vol. 182: L. D. Baumert. Cyclic Difference Sets. VI, 166 pages. 1971. DM 16,–

Vol. 183: Analytic Theory of Differential Equations. Edited by P. F. Hsieh and A. W. J. Stoddart. VI, 225 pages. 1971. DM 20,–

Vol. 184: Symposium on Several Complex Variables, Park City, Utah, 1970. Edited by R. M. Brooks. V, 234 pages. 1971. DM 20,–

Vol. 185: Several Complex Variables II, Maryland 1970. Edited by J. Horváth. III, 287 pages. 1971. DM 24,–

Vol. 186: Recent Trends in Graph Theory. Edited by M. Capobianco/ J. B. Frechen/M. Krolik. VI, 219 pages. 1971. DM 18,–

Vol. 187: H. S. Shapiro, Topics in Approximation Theory. VIII, 275 pages. 1971. DM 22,–

Vol. 188: Symposium on Semantics of Algorithmic Languages. Edited by E. Engeler. VI, 372 pages. 1971. DM 26,–

Vol. 189: A. Weil, Dirichlet Series and Automorphic Forms. V, 164 pages. 1971. DM 16,–

Vol. 190: Martingales. A Report on a Meeting at Oberwolfach, May 17-23, 1970. Edited by H. Dinges. V, 75 pages. 1971. DM 16,–

Vol. 191: Séminaire de Probabilités V. Edited by P. A. Meyer. IV, 372 pages. 1971. DM 26,–

Vol. 192: Proceedings of Liverpool Singularities – Symposium I. Edited by C. T. C. Wall. V, 319 pages. 1971. DM 24,–

Vol. 193: Symposium on the Theory of Numerical Analysis. Edited by J. Ll. Morris. VI, 152 pages. 1971. DM 16,–

Vol. 194: M. Berger, P. Gauduchon et E. Mazet. Le Spectre d'une Variété Riemannienne. VII, 251 pages. 1971. DM 22,–

Vol. 195: Reports of the Midwest Category Seminar V. Edited by J. W. Gray and S. Mac Lane. III, 255 pages. 1971. DM 22,–

Vol. 196: H-spaces – Neuchâtel (Suisse)- Août 1970. Edited by F. Sigrist, V, 156 pages. 1971. DM 16,–

Vol. 197: Manifolds – Amsterdam 1970. Edited by N. H. Kuiper. V, 231 pages. 1971. DM 20,–

Vol. 198: M. Hervé, Analytic and Plurisubharmonic Functions in Finite and Infinite Dimensional Spaces. VI, 90 pages. 1971. DM 16,–

Vol. 199: Ch. J. Mozzochi, On the Pointwise Convergence of Fourier Series. VII, 87 pages. 1971. DM 16,–

Vol. 200: U. Neri, Singular Integrals. VII, 272 pages. 1971. DM 22,–

Vol. 201: J. H. van Lint, Coding Theory. VII, 136 pages. 1971. DM 16,–

Vol. 202: J. Benedetto, Harmonic Analysis on Totally Disconnected Sets. VIII, 261 pages. 1971. DM 22,–

Vol. 203: D. Knutson, Algebraic Spaces. VI, 261 pages. 1971. DM 22,–

Vol. 204: A. Zygmund, Intégrales Singulières. IV, 53 pages. 1971. DM 16,–

Vol. 205: Séminaire Pierre Lelong (Analyse) Année 1970. VI, 243 pages. 1971. DM 20,–

Vol. 206: Symposium on Differential Equations and Dynamical Systems. Edited by D. Chillingworth. XI, 173 pages. 1971. DM 16,–

Vol. 207: L. Bernstein, The Jacobi-Perron Algorithm – Its Theory and Application. IV, 161 pages. 1971. DM 16,–

Vol. 208: A. Grothendieck and J. P. Murre, The Tame Fundamental Group of a Formal Neighbourhood of a Divisor with Normal Crossings on a Scheme. VIII, 133 pages. 1971. DM 16,–

Vol. 209: Proceedings of Liverpool Singularities Symposium II. Edited by C. T. C. Wall. V, 280 pages. 1971. DM 22,–

Vol. 210: M. Eichler, Projective Varieties and Modular Forms. III, 118 pages. 1971. DM 16,–

Vol. 211: Théorie des Matroïdes. Edité par C. P. Bruter. III, 108 pages. 1971. DM 16,–

Vol. 212: B. Scarpellini, Proof Theory and Intuitionistic Systems. VII, 291 pages. 1971. DM 24,–

Vol. 213: H. Hogbe-Nlend, Théorie des Bornologies et Applications. V, 168 pages. 1971. DM 18,–

Vol. 214: M. Smorodinsky, Ergodic Theory, Entropy. V, 64 pages. 1971. DM 16,–

Vol. 215: P. Antonelli, D. Burghelea and P. J. Kahn, The Concordance-Homotopy Groups of Geometric Automorphism Groups. X, 140 pages. 1971. DM 16,-

Vol. 216: H. Maaß, Siegel's Modular Forms and Dirichlet Series. VII, 328 pages. 1971. DM 20,-

Vol. 217: T. J. Jech, Lectures in Set Theory with Particular Emphasis on the Method of Forcing. V, 137 pages. 1971. DM 16,-

Vol. 218: C. P. Schnorr, Zufälligkeit und Wahrscheinlichkeit. IV, 212 Seiten 1971. DM 20,-

Vol. 219: N. L. Alling and N. Greenleaf, Foundations of the Theory of Klein Surfaces. IX, 117 pages. 1971. DM 16,-

Vol. 220: W. A. Coppel, Disconjugacy. V, 148 pages. 1971. DM 16,-

Vol. 221: P. Gabriel und F. Ulmer, Lokal präsentierbare Kategorien. V, 200 Seiten. 1971. DM 16,-

Vol. 222: C. Meghea, Compactification des Espaces Harmoniques. III, 108 pages. 1971. DM 16,-

Vol. 223: U. Felgner, Models of ZF-Set Theory. VI, 173 pages. 1971. DM 16,-

Vol. 224: Revètements Etales et Groupe Fondamental. (SGA 1). Dirigé par A. Grothendieck XXII, 447 pages. 1971. DM 30,-

Vol. 225: Théorie des Intersections et Théorème de Riemann-Roch. (SGA 6). Dirigé par P. Berthelot, A. Grothendieck et L. Illusie. XII, 700 pages. 1971. DM 40,-

Vol. 226: Seminar on Potential Theory, II. Edited by H. Bauer. IV, 170 pages. 1971. DM 18,-

Vol. 227: H. L. Montgomery, Topics in Multiplicative Number Theory. IX, 178 pages. 1971. DM 18,-

Vol. 228: Conference on Applications of Numerical Analysis. Edited by J. Ll. Morris. X, 358 pages. 1971. DM 26,-

Vol. 229: J. Väisälä, Lectures on n-Dimensional Quasiconformal Mappings. XIV, 144 pages. 1971. DM 16,-

Vol. 230: L. Waelbroeck, Topological Vector Spaces and Algebras. VII, 158 pages. 1971. DM 16,-

Vol. 231: H. Reiter, L1-Algebras and Segal Algebras. XI, 113 pages. 1971. DM 16,-

Vol. 232: T. H. Ganelius, Tauberian Remainder Theorems. VI, 75 pages. 1971. DM 16,-

Vol. 233: C. P. Tsokos and W. J. Padgett. Random Integral Equations with Applications to Stochastic Systems. VII, 174 pages. 1971. DM 18,-

Vol. 234: A. Andreotti and W. Stoll. Analytic and Algebraic Dependence of Meromorphic Functions. III, 390 pages. 1971. DM 26,-

Vol. 235: Global Differentiable Dynamics. Edited by O. Hájek, A. J. Lohwater, and R. McCann. X, 140 pages. 1971. DM 16,-

Vol. 236: M. Barr, P. A. Grillet, and D. H. van Osdol. Exact Categories and Categories of Sheaves. VII, 239 pages. 1971, DM 20,-

Vol. 237: B. Stenström. Rings and Modules of Quotients. VII, 136 pages. 1971. DM 16,-

Vol. 238: Der kanonische Modul eines Cohen-Macaulay-Rings. Herausgegeben von Jürgen Herzog und Ernst Kunz. VI, 103 Seiten. 1971. DM 16,-

Vol. 239: L. Illusie, Complexe Cotangent et Déformations I. XV, 355 pages. 1971. DM 26,-

Vol. 240: A. Kerber, Representations of Permutation Groups I. VII, 192 pages. 1971. DM 18,-

Vol. 241: S. Kaneyuki, Homogeneous Bounded Domains and Siegel Domains. V, 89 pages. 1971. DM 16,-

Vol. 242: R. R. Coifman et G. Weiss, Analyse Harmonique Non-Commutative sur Certains Espaces. V, 160 pages. 1971. DM 16,-

Vol. 243: Japan-United States Seminar on Ordinary Differential and Functional Equations. Edited by M. Urabe. VIII, 332 pages. 1971. DM 26,-

Vol. 244: Séminaire Bourbaki - vol. 1970/71. Exposés 382-399. IV, 356 pages. 1971. DM 26,-

Vol. 245: D. E. Cohen, Groups of Cohomological Dimension One. V, 99 pages. 1972. DM 16,-

Vol. 246: Lectures on Rings and Modules. Tulane University Ring and Operator Theory Year, 1970-1971. Volume I. X, 661 pages. 1972. DM 40,-

Vol. 247: Lectures on Operator Algebras. Tulane University Ring and Operator Theory Year, 1970-1971. Volume II. XI, 786 pages. 1972. DM 40,-

Vol. 248: Lectures on the Applications of Sheaves to Ring Theory. Tulane University Ring and Operator Theory Year, 1970-1971. Volume III. VIII, 315 pages. 1971. DM 26,-

Vol. 249: Symposium on Algebraic Topology. Edited by P. J. Hilton. VII, 111 pages. 1971. DM 16,-

Vol. 250: B. Jónsson, Topics in Universal Algebra. VI, 220 pages. 1972. DM 20,-

Vol. 251: The Theory of Arithmetic Functions. Edited by A. A. Gioia and D. L. Goldsmith VI, 287 pages. 1972. DM 24,-

Vol. 252: D. A. Stone, Stratified Polyhedra. IX, 193 pages. 1972. DM 18,-

Vol. 253: V. Komkov, Optimal Control Theory for the Damping of Vibrations of Simple Elastic Systems. V, 240 pages. 1972. DM 20,-

Vol. 254: C. U. Jensen, Les Foncteurs Dérivés de $\varprojlim$ et leurs Applications en Théorie des Modules. V, 103 pages. 1972. DM 16,-

Vol. 255: Conference in Mathematical Logic - London '70. Edited by W. Hodges. VIII, 351 pages. 1972. DM 26,-

Vol. 256: C. A. Berenstein and M. A. Dostal, Analytically Uniform Spaces and their Applications to Convolution Equations. VII, 130 pages. 1972. DM 16,-

Vol. 257: R. B. Holmes, A Course on Optimization and Best Approximation. VIII, 233 pages. 1972. DM 20,-

Vol. 258: Séminaire de Probabilités VI. Edited by P. A. Meyer. VI, 253 pages. 1972. DM 22,-

Vol. 259: N. Moulis, Structures de Fredholm sur les Variétés Hilbertiennes. V, 123 pages. 1972. DM 16,-

Vol. 260: R. Godement and H. Jacquet, Zeta Functions of Simple Algebras. IX, 188 pages. 1972. DM 18,-

Vol. 261: A. Guichardet, Symmetric Hilbert Spaces and Related Topics. V, 197 pages. 1972. DM 18,-

Vol. 262: H. G. Zimmer, Computational Problems, Methods, and Results in Algebraic Number Theory. V, 103 pages. 1972. DM 16,-

Vol. 263: T. Parthasarathy, Selection Theorems and their Applications. VII, 101 pages. 1972. DM 16,-

Vol. 264: W. Messing, The Crystals Associated to Barsotti-Tate Groups: with Applications to Abelian Schemes. III, 190 pages. 1972. DM 18,-

Vol. 265: N. Saavedra Rivano, Catégories Tannakiennes. II, 418 pages. 1972. DM 26,-

Vol. 266: Conference on Harmonic Analysis. Edited by D. Gulick and R. L. Lipsman. VI, 323 pages. 1972. DM 24,-

Vol. 267: Numerische Lösung nichtlinearer partieller Differential- und Integro-Differentialgleichungen. Herausgegeben von R. Ansorge und W. Törnig, VI, 339 Seiten. 1972. DM 26,-

Vol. 268: C. G. Simader, On Dirichlet's Boundary Value Problem. IV, 238 pages. 1972. DM 20,-

Vol. 269: Théorie des Topos et Cohomologie Etale des Schémas. (SGA 4). Dirigé par M. Artin, A. Grothendieck et J. L. Verdier. XIX, 525 pages. 1972. DM 50,-

Vol. 271: J. P. May, The Geometry of Iterated Loop Spaces. IX, 175 pages. 1972. DM 18,-

Vol. 272: K. R. Parthasarathy and K. Schmidt, Positive Definite Kernels, Continuous Tensor Products, and Central Limit Theorems of Probability Theory. VI, 107 pages. 1972. DM 16,-

Vol. 273: U. Seip, Kompakt erzeugte Vektorräume und Analysis. IX, 119 Seiten. 1972. DM 16,-

Vol. 274: Toposes, Algebraic Geometry and Logic. Edited by. F. W. Lawvere. VI, 189 pages. 1972. DM 18,-